Springer Tracts in Modern Physics 104

W0051041

Editor: G. Höhler
Associate Editor: E. A. Niekisch

Editorial Board: S. Flügge H. Haken J. Hamilton
H. Lehmann W. Paul

Springer Tracts in Modern Physics

* denotes a volume which contains a Classified Index starting from Volume 36.

Iven Pockrand

Surface Enhanced Raman Vibrational Studies at Solid/Gas Interfaces

With 60 Figures

Springer-Verlag
Berlin Heidelberg GmbH 1984

Dr. Iven Pockrand

Drägerwerk AG, Postfach 13 39
D-2400 Lübeck 1, Fed. Rep. of Germany

Manuscripts for publication should be addressed to:

Gerhard Höhler

Institut für Theoretische Kernphysik der Universität Karlsruhe
Postfach 6380, D-7500 Karlsruhe 1, Fed. Rep. of Germany

*Proofs and all correspondence concerning papers in the process of publication
should be addressed to:*

Ernst A. Niekisch

Haubourdinstrasse 6, D-5170 Jülich 1, Fed. Rep. of Germany

ISBN 978-3-662-15273-7   ISBN 978-3-540-38952-1 (eBook)
DOI 10.1007/978-3-540-38952-1

Library of Congress Cataloging in Publication Data. Pockrand, Iven, 1943– Surface enhanced Raman vibrational studies at solid/gas interfaces. (Springer tracts in modern physics; 104) Bibliography: p. 1. Raman effect, Surface enhanced. 2. Surfaces (Physics) 3. Surface chemistry. I. Title. II. Series. QC1.S797   vol. 104 [QC454.R36] 539s [530.4'1]   84-5387

This work is subject to copyright. All rights are reserved, whether the whole or part of the material is concerned, specifically those of translation, reprinting, reuse of illustrations, broadcasting, reproduction by photocopying machine or similar means, and storage in data banks. Under § 54 of the German Copyright Law where copies are made for other than private use, a fee is payable to „Verwertungsgesellschaft Wort", Munich.

© by Springer-Verlag Berlin Heidelberg 1984

Originally published by Springer-Verlag Berlin Heidelberg New York Tokyo in 1984
Softcover reprint of the hardcover 1st edition 1984

The use of registered names, trademarks, etc. in this publication does not imply, even in the absence of a specific statement, that such names are exempt from the relevant protective laws and regulations and therefore free for general use.

2153/3130–5 4 3 2 1 0

Preface

Many molecules adsorbed on appropriately prepared metal surfaces display a Raman scattering cross section which is several orders of magnitude greater than the corresponding quantity for the isolated molecule. This effect, surface enhanced Raman scattering (SERS), which was discovered eight years ago, opened the very interesting path to Raman vibrational spectroscopy of sub-monolayer quantities of adsorbates, whose study had formerly been thought to be without prospects because of the insufficient sensitivity of ordinary Raman scattering.

This book comprehensively reviews surface enhanced Raman vibrational studies of solid/gas interfaces. It briefly illuminates the current state of understanding of SERS as inferred from relevant experimental results. Emphasis is put on the presentation and evaluation of SER vibrational data from various molecules adsorbed on metal surfaces, in particular silver and the other noble metals. In addition, applications of SERS to problems in tribology and catalysis as well as related surface enhanced phenomena like enhanced nonlinear optical effects or infrared absorption are described. SER studies of metal electrodes and colloidal suspensions are not treated since these are summarized in several other reviews.

I hope that this volume will be a useful help for surface scientists interested in vibrational spectroscopy of adsorbates and act as a stimulus for future work and progress in the field.

Much of the work presented in the book has been performed during a four year stay at the "Physikalisches Institut III" of the University of Düsseldorf. I would like to thank Prof. A. Otto for the exciting times spent at his institute and many stimulating, fruitful, and critical discussions. The skilful technical assistance of J. Liebetrau in the experimental work performed at Düsseldorf is highly appreciated. I am also indebted to Dr. J. Billmann for a careful reading of the manuscript, to Mrs. B. Derks for the accurate execution of the drawings, and to Mrs. C. Lütjens for the fast and efficient processing of the manuscript.

Many of my colleagues supported the work on this review article by sending information and/or preprints prior to publication. I would like to thank F. Adrian, A. Campion, R. Chang, A. Creighton, J. Demuth, S. Efrima, M. Kerker, P. Liao, H. Metiu, A. Nitzan, M. Philpott, G. Schatz, H. Seki, D. Tevault, J. Tsang, H. Ueba, K. Ushioda, R. Van Duyne, D. Weitz, T. Wood, and H. Yamada.

Finally, I would like to thank my wife Petra and my little daughter Friederike, without whose patience with my almost permanent absence from home during the formation of the review this book would never have been completed.

Lübeck, June 1984   *Iven Pockrand*

Contents

List of Abbreviations

AES	Auger Electron Spectroscopy
AIS	Atom Inelastic Scattering
ATR	Attenuated Total Reflection
CMA	Cylindrical Mirror Analyzer
DIRS	Disorder Induced Raman Scattering
EELS	Electron Energy Loss Spectroscopy
FWHM	Full Width at Half Maximum
IETS	Inelástic Electron Tunneling Spectroscopy
IRAS	Infra - Red Absorption Spectroscopy
IRTS	Infra - Red Transmission Spectroscopy
LEED	Low Energy Electron Spectroscopy
ML	Mono-Layer
NIS	Neutron Inelastic Scattering
OMA	Optical Multichannel Analyzer
SER	Surface Enhanced Raman
SERS	Surface Enhanced Raman Scattering
TDS	Thermal Desorption Spectroscopy
UHV	Ultra High Vacuum
UPS	UV - Photoemission Spectroscopy
XPS	X - Ray Photoemission Spectroscopy
$\Delta\phi$	Work function
L	Langmuir $(1\ L = 10^{-6}\ \text{Torr s})$
D	Debye $(1\ D = 10^{-18}\ \text{esu})$

1. Introduction

Vibrational spectroscopy has been employed for many years to study the structure and bonding of molecules. As each bond has its own, characteristic frequency /1-4/ vibrational spectra and molecular structure are related. Infrared absorption and Raman arrangements have most frequently been used for vibrational studies /5-8/. Other techniques like inelastic scattering of electrons /9,10/, neutrons /11,12/, or atoms /10/, which require more refined experimental set-ups, have found considerably less broad spreading as analytical tools.

Molecules are usually perturbed upon adsorption on solid surfaces. Bond strengths and/or structure may change, new species may be formed due to dissociative adsorption or surface promoted reactions between different adsorbed molecules. Surface vibrational spectroscopy can provide significant information on these changes. Character and concentration of the adsorbed species as well as adsorption geometry and site might be extracted from the data. To facilitate evaluation, vibrational spectroscopy is usually backed by other surface sensitive techniques like, e.g., ultraviolet photoemission spectroscopy (UPS), low energy electron diffraction (LEED), thermal desorption spectroscopy (TDS), or work function measurements ($\Delta\Phi$).

Several experimental techniques have been developed to study vibrations of adsorbed molecules. Infrared transmission (IRTS) or infrared reflection absorption spectroscopy (IRAS) /13,14/ and electron energy loss spectroscopy (EELS) /15,16/ have found widespread popularity. As outlined in detail in a recently published book /17/, neutron and atom inelastic scattering (NIS and AIS) as well as inelastic electron tunneling spectroscopy (IETS) /18/ are also becoming established as useful, surface sensitive techniques. Raman spectroscopy, however, did not attract the attention of surface scientists (until recently, see below), although this method combines several advantages in a unique way. A resolution of 1 cm^{-1} and a free spectral range of 100 - 4000 cm^{-1} are easily obtainable, solid/gas surfaces under high pressure or solid/electrolyte interfaces may be investigated in situ, and, finally, one may get additional information from depolarization measurements. However, the sensitivity of Raman scattering is poor and has been believed to be in general insufficient for vibrational studies of adsorbed molecules. To illustrate this fact let us estimate the intensity which is inelastically scattered from a

monolayer of molecules adsorbed on a perfectly reflecting metal surface. The detected Raman intensity I_{Raman} is given by /19/:

$$I_{Raman} = 4\Omega \ (d\sigma/d\Omega)nNAQT_sT_0 \qquad (1)$$

where $(d\sigma/d\Omega)$ is the differential Raman cross section, Ω the collected solid angle,
n the flux of incident photons, N the density of adsorbed molecules, A the illuminated area, Q the quantum efficiency of the detector, and T_s and T_0 are the transmission of the spectrometer and the collecting optics, respectively. The factor
four considers the influence of the perfectly reflecting metal on the incident as
well as on the Raman scattered photons. For a Lorentzian line shape of the vibrational mode (halfwidth Γ), the integrated intensity I_{Raman} can be converted into a
peak intensity I_{Raman}^{peak} by /19/:

$$I_{Raman}^{peak} = I_{Raman}/(\pi\Gamma) \ . \qquad (2)$$

For the specific case of adsorbed pyridine C_5H_5N one has $N = 5 \cdot 10^{14}$ molecules/cm^2
/20/ and a relatively large (gas phase) Raman cross section of $3.3 \cdot 10^{-30}$ cm^2/(sterad
molecule) for the symmetric breathing vibration /21,22/. Using 200 mW of 514.5 nm
Ar-ion laser radiation focused down to $A = 3 \cdot 10^{-2}$ cm^2, and assuming $\Omega = 1$ sterad,
$Q = 0.15$, $T_s \cdot T_0 = 5 \cdot 10^{-3}$, and a photon counting system which records all multiplier
pulses, one expects an integrated Raman intensity of at best 15 cts/s. This gives
a peak intensity of ≈ 2 cts/s ($\Gamma = 4$ cm^{-1}, spectrometer bandpass 2 cm^{-1}), which is
not particularly encouraging if one thinks of surface vibrational studies. Indeed,
early investigations yielded Raman signals only from strong Raman scatterers on
high surface area adsorbents /23/ or from relatively thick films (≈ 5 nm) of strong
scatterers on silver films /24/. Thinner overlayers may be detected, if the background intensity in the Raman spectra is sufficiently low. This has been shown for
pyridine on Ag(110) /25/ and Ni(111) /26/. Figure 1a displays the Raman spectrum of
a sample coated with roughly three layers of pyridine. A peak intensity of ≈ 10 cts/s
in rough agreement with the expected value has been measured. When using a more
elaborate experimental arrangement, Raman spectra from less than a monolayer of adsorbed molecules on Ni(111) /27/ or Ag(111) (Fig. 1b, /28/) can be taken. These results open the very interesting path to Raman vibrational studies of adsorbates on
well characterized single crystal surfaces (with, however, still moderate sensitivity if compared to, e.g., EELS).

As seen from (1), I_{Raman} can be increased when the flux n or, correspondingly,
the electromagnetic field strength at the site of the Raman scatterer is increased.
This has been accomplished in the early seventies for thin organic layers by incor-
porating these films into suitable, layered structures /29-32/, so that guided light
modes in the film or plasmon surface polaritons at the film/metal interface can

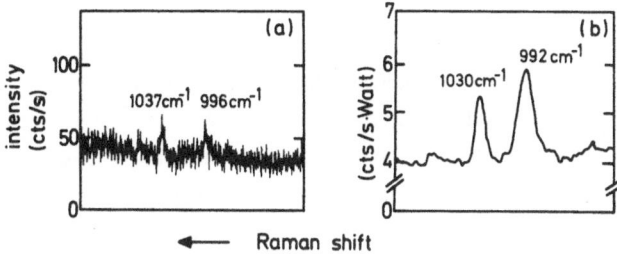

Fig. 1. Ordinary Raman spectra from pyridine on silver. a) ≈ 3 layers on Ag(110), $T_s = 150$ K; 250 mW of 514.5 nm radiation, 2 cm^{-1} bandpass; b) ≈ 1 monolayer on Ag(111), $T_s = 110$ K; 1000 mW of 514.5 nm radiation, 10 cm^{-1} bandpass (after /28/). Symmetric (992/996 cm^{-1}) and antisymmetric breathing mode (1030/1037 cm^{-1})

propagate (these optical modes are discussed in, e.g., /33/). Raman spectra of good quality have been recorded, when a thin film or interface mode was resonantly excited. The influence of long wavelength, extended surface plasmon polaritons on the Raman scattering from molecules on highly reflecting metal surfaces (e.g. Ag) has subsequently been investigated in some more detail. Besides the resonant enhancement of the incident field the calculations generally consider also the resonant emission of the Raman scattered photons via plasmon surface polaritons. For the attenuated total reflection (ATR) configuration /34-36/ as well as for grating surfaces /37-41/ enhancements of the Raman scattered intensity of $10^3 - 10^6$ have been calculated under favourable conditions (the ATR technique is described in, e.g., /33/, plasmon surface polaritons on gratings are treated in, e.g., /42/). Experimentally observed enhancements are usually much smaller, between ≈ 5 and ≈ 100 (/43-48/; only recently a rather large factor of $\approx 4 \cdot 10^4$ has been estimated from an ATR-Raman study /49/. Grating surfaces have especially been used in tunnel junction structures /50-54/). Nevertheless, the enhancement brought about by excitation of surface waves may render possible or facilitate surface Raman vibrational studies in certain cases. As this technique is only applicable to molecules adsorbed on certain materials with appropriately corrugated surface or incorporated into an ATR configuration, it did not find, however, widespread attention and interest in the community of surface scientists.

This was different, when two research groups independently reported a giant enhancement ($10^5 - 10^6$) of the Raman cross section of pyridine on silver electrodes /55,56/ (actually, similar Raman spectra from pyridine on Ag electrodes had been published earlier /57-59/; however, these authors did not realize the unusual enhancement). Enhanced Raman signals were only observed after a proper activation of the electrode by an oxidation-reduction cycle. Soon after the first report of surface enhanced Raman scattering (SERS) from silver electrodes SERS was also observed from molecules on silver colloids /60/ and on silver/gas (vacuum) interfaces /61/. It became apparent, that the surfaces of only certain metals were SERS "active"

3

(group Ib mainly), which had to be pretreated appropriately or prepared under special conditions (the significance of this SERS "activity" will be outlined later). Moreover, not all, although many, molecules adsorbed on SERS active surfaces displayed enhanced Raman scattering equally well.

Several mechanisms have been proposed to contribute to SERS (including excitation of extended plasmon surface polaritons as mentioned above). The various models have been extensively discussed in several review articles /62-68/ and a recently published book on SERS /69/. Therefore they will be only briefly exposed in Chapt. 2 of this volume. Some basic experimental facts and the theoretical concepts, which are presently accepted by most groups active in the field, will as well be summarized in this chapter.

The experimental situation is less comprehensively reviewed. Early experimental results from solid/electrolyte interfaces are contrasted with theoretical concepts in /19,70/, applications of Raman spectroscopy in electrochemistry are discussed in /71/, and some experimental observations from solid/electrolyte as well as solid/gas interfaces are listed in /72/. Several recent articles /73-76/ discuss selected, relevant observations in connection with actual theoretical developments. This review summarizes SER experimental studies from solid/gas interfaces. The impact of experimental facts on the theoretical discussion will be displayed, but no effort is made to comprehensively appraise theoretical concepts (for this the interested reader is referred to /66-69/). Rather, surface enhanced Raman vibrational spectra from various molecules on metal surfaces will be evaluated in some detail. Therefore the volume addresses also the surface scientist, who is not particularly interested in the details of the theoretical discussion, but rather wants to be informed of the applications and potential of surface enhanced Raman scattering as a surface analytical tool.

The paper is organized as follows. After a brief survey of basic experimental facts, proposed models, and the present state of the theory discussion (Chapt. 2), some details of the experimental techniques will be illuminated in Chapt. 3. A rather detailed analysis of SER data from pyridine on metals will be presented in Chapt. 4. The relevance of some results with respect to theoretical concepts will be accentuated. SER vibrational spectra from hydrocarbons, carbon monoxide and carbonaceous species, oxygen, and water adsorbed to metals are discussed in Chapts. 5-8. Data from other, less fully investigated adsorbate/metal systems are summarized in Chapt. 9. Relative broad room will be given to results from molecules on "coldly" evaporated films, since these usually display the most detailed spectra ("coldly" evaporated films are characterized in Chapt. 3). In Chapt. 10 some applications of SERS to problems in, e.g., tribology or catalytic activity of metal surfaces are presented. Finally, momentary problems and state of the art are reflected in Chapt. 11. In an outlook, future capabilities and limitations of Raman spectroscopy as a surface analytical tool are displayed.

This article almost totally ignores the very interesting and important SERS work on metal electrodes and colloidal suspensions. The reader, who is also interested in these aspects of SERS, is referred to another review /77/.

2. Fundamentals of Surface Enhanced Raman Scattering

In this chapter we give an overview on the present experimental as well as theoretical situation of SERS. No attempt is made to exhaustively quote all related
work, and I apologize to those, whose work did not find the attention it deserves.

A few points require special comments. Firstly, much confusion has been intro-
duced into the field by experimental papers, whose results or interpretations were
not carefully enough cross checked. I shall express scepticism, whenever it is
necessary, i.e. when results could not be reproduced. Secondly, a variety of observations has been classified as SERS, often without elaborating the specific properties of the system under investigation. The "giant" enhancement (10^5 - 10^6) at
appropriately pretreated silver electrodes /55,56/, the weak effect (enhancement
≈ 5 - 100) when resonantly exciting plasmon surface polaritons at optical gratings
/46/ or in an ATR configuration /43,44/, as well as Raman spectra from adsorbed
molecules on for instance silica supported Ni catalysts /78/ have all been simply
labeled SERS. To the outsider not familiar with the field this may have suggested
one single enhancement mechanism similarly working in quite different systems
(which is a wrong picture). Thirdly, theoretical concepts developed for special
configurations like gratings or isolated metal spheres have sometimes been intro-
duced so, as if they were capable to explain all or almost all aspects of SERS in
every system. Unfortunately, the situation is more complicated. Finally, I would
like to remind of the "pre SERS" Raman work on adsorbed molecules. Numerous papers
report on (ordinary) Raman studies of molecules physi- or chemisorbed on Ni single
crystal surfaces (/79/, see also /27/), oxide surfaces like silica or alumina /80,
81/, supported metal catalysts /82/, or metal electrodes /83,84/. Several review
articles summarize these investigations /59,85,86/. The important messages from
these studies are: (i) only an extremely careful quantitative evaluation of scattered intensities allows to safely decide, whether an observed weak Raman signal
is surface enhanced or not, and (ii) laser Raman spectroscopy - ordinary or enhanced - can provide valuable information on adsorbed molecules.

2.1 Basic Experimental Observations

As was already evident from the very first SER studies at silver electrodes, only
samples activated by an anodic oxidation-reduction cycle exhibited strongly enhanced
Raman signals from adsorbed molecules (/55-57/; enhancement factor 10^5 - 10^6). The
pretreatment (activation) has been shown /87,88/ to change the surface topography:
SERS active silver electrodes are rough. The roughness scale "important" for SERS
is still a matter of debate /64,67,89/. There is, however, agreement that some kind
of roughness is a necessary prerequisite for SERS /64/.

   SERS is not restricted to silver electrodes. Enhanced Raman signals have been
reported for several other activated silver interfaces or specially prepared systems:

- mechanically polished, polycrystalline silver sheets measured in air (/90/; no
  quantitative estimation of the enhancement factor for cyanide deposited by immersion in alkaline KCN solution)
- silver island films vapour deposited on glass (/62,91/; enhancement factors of
  $\approx 10^5$ have been observed for adsorbed isonicotinic acid /92/, p-nitrobenzoic
  acid /93/, and pyridine /94/)
- silver aqueous sol particles of dimensions comparable to or less than the wavelength of light (/60/; for citrate ions adsorbed to silver particles of 42 nm
  diameter an enhancement factor of $6 \cdot 10^5$ has been measured /95/)
- silver optical gratings with periods comparable to the wavelength (/46/; a weak
  enhancement of ≈ 30 for thin polystyrene coatings /47/ and of $\approx 10^2$ for pyridine
  /48/ due to resonant excitation of plasmon surface polaritons has been observed)
- polycrystalline silver foils cleaned and probably roughened by Ar-ion bombard-
  ment in UHV (/96,97/; for adsorbed pyridine enhancement factors of 10^3 - 10^5 /96/
  and $\gtrsim 10^3$ /97/ have been estimated, where the first value is uncertain because
  of difficulties in measuring the dosing rate)
- photochemically roughened silver surfaces with roughness features of typically
  50 nm lateral extensions (/98/; an enhancement factor of $\approx 5 \cdot 10^4$ for pyridine
  has been observed)
- coldly evaporated silver films, i.e. thick silver films evaporated on substrates
  cooled to typically 120 K (/61/; Raman signals from adsorbed pyridine display
  an enhancement of $\approx 10^4$ /99/)
- Al-Al$_2$O$_3$-Ag tunnel junctions evaporated on rough CaF$_2$ films or on optical gratings (/51/; for 4-pyridine-carboxaldehyde at the Al$_2$O$_3$-Ag interface an enhancement of $\lesssim 20$ for junctions on gratings has been estimated due to resonant excitation of plasmon surface polaritons; CaF$_2$ roughened structures have not been
  evaluated quantitatively).

The enhancement factors given above have usually been determined for the
strongest line of the adsorbate (for pyridine, this is the symmetric ring breathing
vibration). The quality of measured spectra in terms of peak intensities and signal

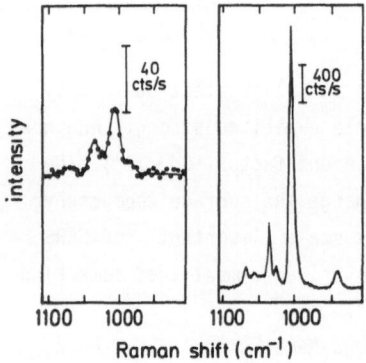

Fig. 2. SER signals from 0.1 monolayer of pyridine on Ag. Left: photochemically rough- ened surface (488 nm radiation, 8 cm^{-1} band- pass; after /98/). Right: coldly evaporated film (170 mW of 514.5 nm radiation, 3 cm^{-1} bandpass; after /100/). For both cases an en- hancement factor of $\approx 10^4$ has been estimated

to noise ratio is rather different for different systems, even when comparable en- hancement factors are estimated. The spectra displayed in Fig. 2 have been recorded under similar experimental conditions. They show surface enhanced Raman signals from roughly a tenth of a monolayer of pyridine on photochemically roughened silver /98/ and on coldly evaporated silver films /100/. Either signal has been estimated to be about four orders of magnitude enhanced. Figure 2 might indicate a too optimistic valuation of enhancement factors in some cases.

Well prepared, smooth single crystal surfaces do not enhance the Raman signal from adsorbed molecules beyond that what is expected from Fresnel equations (/25,28/, pyridine on Ag). A weak enhancement of $\approx 4\cdot10^2$ reported for pyridine on Ag(100) /101/ might be the combined result of residual roughness as outlined in /28/, of the flat

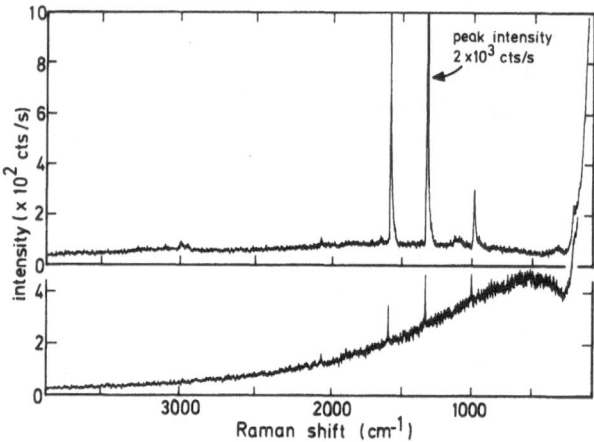

Fig. 3. SER spectra from coldly evaporated Ag films exposed to 3 L of ethylene. Upper trace: sample exposed and measured at 120 K. Lower trace: sample annealed to 260 K ($\approx$ 1 K/min), recooled to 120 K, exposed and measured. 200 mW of 514.5 nm ra- diation, 4.5 cm^{-1} bandpass. After /109/

metal surface contribution given by the Fresnel equations /102/, and of uncertainties in the evaluation procedure.

SER lines from adsorbates are accompanied by a continuous background scattering
which extends beyond 4000 cm^{-1} /90,103,104/. The background is also observed from
SERS active surfaces *without* adsorbed molecules /105 - 108/. Hence it is an intrinsic
property of the metal. For silver electrodes, both, background and SER lines, increase with increasing activation /104/. Therefore the two phenomena may have important features in common, as assumed in /106,107/. The correlation of background
and SER intensity is, however, not always observed as shown in Fig. 3 for ethylene
on coldly evaporated silver films /109/. The background has been interpreted as
luminescence /110,111/ due to roughness assisted, radiative decay of electronic excitations from a continuum of states /112,113/.

Ordinary Raman selection rules are relaxed in SERS. IR allowed vibrations of centrosymmetric molecules, which are subject to the principle of mutual exclusion, have
been detected in SER spectra (for instance pyrazine on silver electrodes /114,115/).
Even silent modes were observed with appreciable intensity (e.g., benzene on silver
films /116/). The breakdown of selection rules has been attributed to the change of
symmetry upon adsorption /114,116/ or, alternatively, to the large electric field
gradients which exist near a metal surface /116-118/. In /67/ yet another explanation based on the "charge transfer picture of the chemical contribution to SERS"
(see below) is given.

Relative SER line intensities differ in general from the corresponding values
of ordinary scattering from the isolated molecule, i.e. mode selective enhancement
is observed /55/. The relative SER intensities vary with electrode potentials /19,
88/ and with the wavelength of the exciting light /19,119/. They are different for
the same molecule adsorbed to different, SERS active metal substrates /120 - 123/,
and they are different for differently prepared SERS active surfaces of the same
metal. The most striking example for the latter is the SER signal of the C-H stretching vibrations of pyridine and of other molecules on silver. It is comparable in
intensity to the strong breathing mode signal for activated electrodes /55/, whereas it is roughly two orders of magnitude smaller for coldly evaporated films (/108/;
see also Chapt. 4).

Overtones and combination bands are absent or only weakly pronounced /124/. SER
lines are depolarized, even if the corresponding lines of the isolated molecule are
strongly polarized /64/.

SER excitation spectra do not show *sharp* resonances. Only slow variations or
broad maxima have been observed. Excitation profiles are different for electrodes
/19,46,119,125 - 127/, for colloidal dispersions /60,95,128 - 131/ and matrix-isolated
particles /132,133/, and for vapour-deposited films in vacuum /94,119,123,134 - 136/
and island films /93,137,138/. Excitation spectra and their impact on theoretical
concepts will be discussed in some detail in Chapt. 4.

Besides silver, which is still the most widely used material for SER studies, appropriately prepared Cu and Au surfaces are SERS active as well under red light illumination (/123,139,140/; SER signals disappear for excitation wavelengths $\lesssim 570$ nm /123/). Other highly reflecting materials like the alkali metals lithium /133/, potassium /141/, and sodium /142/ exhibit also SERS. Preliminary results from Al films /143,144/ require further careful experiments to establish the degree of the enhancement (very recently, a weak enhancement of $\approx 10^3$ has been reported for p-nitrobenzoic acid on aluminum particle arrays /145/). Alloying gold to silver quenches the SER signal for green/blue light excitation, but not for red /146/; $\gtrsim 5\%$ Pd in Ag also quenches the enhancement below the limit of detection /147,148/.

Other reports of SERS from low reflectivity metals like Hg /149,150/, Cd /151/, Pd /152 - 154/, Pt /152 - 157/, Ti /154/, and Ni /153,154,157 - 161/ should be valu- ated very critically. Some results could not be reproduced by other groups (e.g. Hg, Cd /162/), and some might be interpreted in terms of ordinary resonance or pre-res- onance Raman scattering rather than in terms of SERS (J_2 on Pt, Pd /152/). Because of the high area surface of supported metal catalysts the relatively weak signals from Ni, Pt /156,158 - 161/ might be just ordinary Raman scattering (independent of the interpretation these Raman vibrational studies of supported metal catalysts yielded interesting results).

Finally, SERS has been reported for pyridine on metal oxides (NiO /152/, TiO_2 /154/), for iodine on a semiconductor electrode (TiO_2, /163/), and for molecular oxygen on an organic single crystal (polydiacetylene, /164/). The latter has been attributed to adsorption induced resonance Raman scattering. Molecular oxygen forms a complex with polydiacetylene with a well defined electronic transition at 2.39 eV /164/. It is interesting within the context of the present discussion of SERS mech- anisms (/67/ and Sect. 2.2) that this transition involves a significant degree of charge transfer. The former two results as well as the studies on low reflectivity metals require more careful experimental work to unambiguously clear the situation.

There seems to be no limitation in molecules which exhibit SERS. Enhanced Raman signals have been observed from simple adsorbates like, e.g., halide ions /165/ and complicated molecules like pyridine derivatives or nucleic acid components (/166, 167/; all on silver electrodes). However, the magnitude of the enhancement might be different for different molecules on the same surface as has recently been demon- strated for coadsorbed CO and N_2 /133/.

There is a "first layer effect" in surface enhanced Raman scattering: molecules in the first layer often show a much stronger enhancement than those in consecutive layers /99,136/. The effect might be restricted to specifically adsorbed molecules in the first layer as is assumed within the concept of SERS active sites /67,87/.

2.2 Theoretical Concepts

The extensive theoretical work on surface enhanced Raman scattering is summarized in several reviews. A survey of early concepts is presented in /63/, more recent work is briefly illuminated in /102/. A detailed, critical valuation of some SER models can be found in /66/ (electromagnetic effects at various SERS active surfaces), in /67/ ("electromagnetic" and "chemical" contributions to SERS), and in /68/.

The inelastically scattered intensity I_{Raman} for an isolated molecule may be written as /168/:

$$I_{Raman} \sim \omega_s^4 \cdot |\alpha|^2 \cdot F^2 \quad . \tag{3}$$

Here ω_s is the Stokes frequency, F the electric field strength of the incident radiation, and α a component of the Raman tensor /168/ (phenomenologically, α describes the normal coordinate derivative of the polarizability of the molecule). I_{Raman} is usually very small, much smaller than the elastically scattered Rayleigh intensity /168/. It may be up to six orders of magnitude stronger when the incident frequency is in resonance with a real transition of the molecule (resonance Raman effect, see, e.g., /169/). Upon adsorption on a metal surface, the electronic states of a molecule which shows only ordinary scattering may be perturbed such as to allow for resonance Raman scattering. This "adsorption induced resonance Raman effect" led to the prediction of enhanced Raman scattering from molecules near a metal surface in /170/, which has been published before the discovery of SERS (in /170/ interaction between excited molecular states and surface plasmon modes is thought to perturb the molecule). An "effective" Raman tensor α_{eff} may take into account such effects. More general, for an adsorbed molecule (3) has to be modified to

$$I_{Raman} \sim \omega_s^4 \cdot |\alpha_{eff}|^2 \cdot (F^2 \cdot G_L) \cdot G_S \quad . \tag{4}$$

Now α_{eff} contains any change of the molecular polarizability upon adsorption or, more exactly, describes the polarizability derivative of the adsorbate/adsorbent system. G_L and G_S account for "electromagnetic" effects: the electric field at the site of the adsorbed molecule and the Stokes emission might both be amplified by the presence of the metal.

I shall first discuss electromagnetic effects. These are small on flat metal surfaces. Due to interference effects, the local field as well as the Stokes emission field might each have up to twice the value of the corresponding quantity for the isolated molecule /171/. Since the local field is almost perpendicular to the surface, "selection rules" appear as outlined in /102/ (see also /172/). This might allow to determine adsorption geometries (similarly, EELS (dipole scattering) and IRAS are governed by selection rules /17/).

The evanescent fields of extended plasmon surface polaritons /33/ can give rise to stronger electromagnetic effects. As mentioned in Chapt. 1, these interface modes can resonantly be excited in an ATR configuration /33/ or at a grating surface /42/. For the former, a resonance enhancement of the intensity of the local electric field at a silver surface of $\lesssim 250$ /34/ and an enhanced Stokes emission on the prism side of $\lesssim 300$ /36/ has been calculated. This gives an overall enhancement of $\lesssim 7 \cdot 10^4$. For optical silver gratings, the corresponding values are $\lesssim 10^4$ for G_L /39/ and $\lesssim 5 \cdot 10^2$ for G_S /173/ resulting in a total enhancement of $\lesssim 5 \cdot 10^6$. Taking into account radiative damping of surface polaritons, a considerably smaller G_L of ≈ 25 has been calculated (/174/; G_S should be affected similarly). As is obvious from a comparison to measurements (see Chapt. 1), most calculations overestimate the plasmon surface polariton contribution to the enhancement.

Similarly, Raman scattering from molecules on isolated metal particles (e.g. on diluted colloidal dispersions) is enhanced by electromagnetic resonances (localized surface plasmons). Calculations within the Rayleigh-small-particle limit have been performed for spheres /175 - 178/ and spheroids /179/, rigorous electrodynamic calculations for spheres /180,181/ and, very recently, for prolate ellipsoids /182/. Numerical evaluations generally assume bulk optical properties for the small metal particles. Only in /183/ the size dependence of the imaginary part of the dielectric function due to surface scattering has been taken into account. Dielectric environment effects have been studied in /184,185/. Depending on the shape and the dimensions of the particle, total enhancement factors of $\approx 10^2 - 10^6$ have been calculated for silver particles.

If a rough metal surface is modeled by an ensemble of isolated hemispheroids protruding from a perfectly conducting plane, the same formalism as for isolated spheroids may be used to calculate enhancement factors (/186 - 188/; note, that only a perpendicular resonance exists for this configuration because of the image effect of the ideally conducting plane). For sharp, needle-like protrusions enhancement factors up to 10^{11} (!) for molecules on the tip of the structure have been calculated due to resonant excitation of surface plasmons and the lightning rod effect, i.e. the concentration of the electric field at parts of the surface with extreme curvature. A slightly more realistic case has been treated in /189/. Here electromagnetic resonances of an isolated hemispherical bump protruding from the plane boundary of a halfspace with the same dielectric function as the bump material have been studied. Numerical results are presented for a particular position of the scattering molecule only.

The calculations mentioned so far neglect interaction between the metal particles or the bumps on the surface. Because of the long range of the elctromagnetic fields of the resonance, this is usually a crude approach. Collective interactions have been treated with Maxwell-Garnett theory /136,190,191/. Within this frame, the optical properties of metal spheres in a dielectric matrix are described by an effective dielectric function, which contains - besides the dielectric function of the metal and the environment - only the volume fraction (filling factor) of the metal /192,193/. A rough, bumpy surface is modeled by a transition layer, whose electromagnetic resonance is then given by the Maxwell-Garnett approach /190/. No absolute numerical results have been presented. For the simple case of two metal spheres, electromagnetic interaction leads to the appearance of two resonances, whose splitting depends on the interparticle distance, and to a substantial enhancement of the field between the spheres /194/.

In a different approach the dipole moments induced in the metal particles or bumps are treated as point dipoles. Dipole-dipole coupling between randomly distributed *small* particles in a dielectric host broadens the electromagnetic resonance and shifts it to the red (with respect to the Maxwell-Garnett result; /195/). The broadening leads to a decrease of G_L as well as of G_S. As shown in /196/, the transverse collective electromagnetic resonance of a square array of uniformly shaped oblate ellipsoids on glass gives a total enhancement factor of $\approx 3 \cdot 10^6$ for molecules uniformly adsorbed on the ellipsoids. In this case, which is regarded as representative for an island film, the contribution of the image dipoles to the total field has also been taken into account. It has been pointed out /196/, that the transverse resonance will be inhomogeneously broadened due to randomly distributed sizes, shapes, orientations, and spacings of the islands in an actual evaporated film, which may reduce the enhancement by two orders of magnitude. The calculations have recently been extended to spheroids of any shape in ordered square lattices or on random positions, and to square arrays of spheroids of random shape /197/. Effects of retardation, radiative damping, as well as finite size contributions to the dielectric response of the island film were discussed. An intensity enhancement (G_L) of 1 - 2 orders of magnitude was estimated.

A real, rough metal surface may be described by a random distribution of metal hemispheroids on a perfectly conducting flat plane /198,199/. If dipolar coupling between the protrusions is taken into account, G_L is calculated to $\approx 10^2$ for Ag (/198/; average over the whole surface). *Small* scale, randomly distributed roughness may be treated with first order perturbation theory (Born approximation) as has been done for elastic Rayleigh scattering /200 - 203/. The approach breaks down for $\delta \approx$ 3 nm, where δ is the rms-value of the roughness amplitude /204/. The roughness induced increase of the radiation from an oscillating dipole relative to the flat surface has been estimated to ≈ 10 within this model /204/ (silver; $\delta = 3$ nm; $\sigma =$ 2 nm, which is at the limit of validity of the Born approximation; σ: correlation length). With a different approach somewhat larger enhancement factors $(10^2 - 10^3)$ have recently been calculated (/205/; $\delta = 15$ nm, $\sigma = 40$ nm).

Several other approaches /35,93,206/ to quantitatively estimate the effect of roughness on excitation and emission of a Raman dipole on a metal surface are treated in /67/. Here the interested reader will find a comprehensive, critical discussion of proposed models for the electromagnetic enhancement.

Let us now briefly touch the "molecular enhancement" mechanisms, which are contained in the effective polarizability α_{eff} of (4). The "image field" model /171, 207 - 213/ considers the influence of the image field on the adsorbate polarizability (in /213/, the influence of other adsorbed molecules and their images on the field at a given molecule has also been taken into account). The adsorbed molecule is usually treated as a point dipole located at a certain distance R from a sharp metal boundary. The effective polarizability derivative α_{eff} of the system (dipole plus image dipole) then varies rapidly with the metal-adsorbate separation and may be highly peaked within a small interval of distances. For silver, an enhancement factor of $\approx 10^7$ for R = 1.41 Å has been calculated, which drops by more than three orders of magnitude when moving the molecule by only 0.1 Å to R = 1.50 Å /211/. More realistic, refined image field models /206,214 - 218/, which take into account the finite molecular size, (and/or) spatial dispersion of the metal dielectric response, (and/or) the continuous variation of the electron density across the interface, or use a coupled-state quantum formalism /219 - 221/, yield considerably smaller enhancement factors. The status is at present still uncertain, since different groups estimate enhancement factors of ≈ 1 /216/ and 10^4 /214/ for apparently similar systems and approaches /212/.

A second group of models considers the interaction of the vibrating ion cores of the adsorbed molecule with the electrons of the metal /39,112,132,222 - 226/. All are based on the idea, that participation of the highly polarizable metal electrons in the Raman process may enhance the cross sections. The ion cores may interact with the metal electrons via coulomb forces /112/ and thus modulate the electronic polarizability at the surface giving rise to so called "Raman reflection" /222,223/. For chemisorbed molecules this mechanism may be accompanied by vibrationally modulated charge transfer to and from the metal into the molecule /132,224,225/, which also modulates the electronic polarizability. Yet another mechanism is investigated in /39/. Here it is assumed, that the motion of the molecular ionic charges modulates the surface barrier potential for tunneling of metal electrons to the molecular site. This, on the other hand, modulates the surface charge density induced by the exciting field, which results in the emission of Raman Stokes photons. Enhancement factors of 10 - 100 /102/ due to vibrational modulation of metal electrons have been estimated for *flat* surfaces.

Finally, we briefly touch models which may be summarized under "adsorption induced resonance Raman scattering". Within this frame it is assumed, that (i) the electronic states of the molecule are perturbed by interaction with the metal, and/or (ii) an additional transition from metal states below the Fermi level to the lowest unoccupied molecular level becomes possible so as to allow for ordinary resonant Raman scattering. Early papers /170,209,227,228/ focused on the coulomb interaction between molecule and metal. This process is of long range, i.e. not restricted to the first layer of adsorbed molecules. Shift and broadening of the molecular levels were estimated by using a formalism, which was developed to describe the properties of an oscillating dipole close to a metal surface /229/. Finite molecular size and nonlocal metal response have been included in a recent treatment of vibrational properties of diatomic molecules on metals /230/. Another, rather special mechanism - formation of a surface complex upon adsorption with a new optical transition in the visible /126/ - gives enhanced Raman scattering only for molecules in direct contact with the metal. The same holds for the equally rather special situation discussed in /231/. The importance of charge transfer excitations for SERS /65, 232 - 238/ (case (ii) from above) has recently been discussed in detail /67/. This mechanism requires chemisorption of the adsorbate, i.e. is a short range effect. Along with other processes mentioned earlier /132,224,225/, which also involve charge transfer from and to the metal, it is usually called the "chemical" contribution to SERS. The role of charge transfer excitations in SERS may best be illustrated by the approach of /236/. Here it is assumed that the lowest unoccupied level of the molecule is broadened to a resonance upon chemisorption due to partial filling of this orbital by metal electrons. The transition of electrons from metal states below the Fermi energy to about the maximum in the density of states of the molecular resonance gives rise to a weak resonance in the Raman cross section. Enhancement factors of ≈ 50 for a "typical case" of chemisorption on silver have been estimated /236/.

The magnitude of any chemical contribution to SERS depends most likely on adsorption geometry and environment for a given molecule/metal system (experimental evidence for this is discussed in /67/). Within the concept of SERS active sites /64, 87, 239/ it has been proposed, that the chemical effect is particularly strong, if the molecule is adsorbed to sites of atomic scale roughness (this concept is often also addressed as the "adatom model" /87/). Adsorption induced resonance Raman models, the contribution of local electronic excitations to SERS, and the role of adatoms have independently also been discussed by a Russian group /226,240 - 245/. For several other interpretations of the enhancement mechanism, which have been developed to explain experimental results from rather special systems, the interested reader is referred to the original literature /246 - 250/. Finally, we mention the microscopic approaches to SERS of /251 - 253/ (ab initio Hartree-Fock cluster calculations) and of /219 - 221,254/ (coupled molecule-surface plasmon formalism).

Various proposed mechanisms and calculated enhancements are summarized and divided into four classes in Table 1. "Electromagnetic" (or "classical") mechanisms are usually of long range, i.e. not restricted to the first layer of adsorbed molecules. Surface corrugation (roughness) is necessary, except in the ATR configuration. Their magnitude depends on the dielectric properties of the metal, but they should work equally well for all adsorbates. The listed "field" effects are only important for small metal-molecule separation (say $\lesssim 1$ nm). They do not need surface roughness.

Table 1. *Proposed* enhancement mechanisms. Enhancement factors E, G_L, G_S have been calculated in the quoted articles. Mechanisms placed on the separation between two classes contain elements of either class (further explanations in the text)

"Electromagnetic" ("Classical") mechanisms	"Field" mechanisms	"Chemical" mechanisms	"Orchid" mechanisms
Local field and Stokes emission enhancement by plasmon type resonances	Enhancement of system polarizability by field mediated metal-molecule interaction	Enhancement of system polarizability by metal-molecule interaction involving charge transfer	
extended plasmon surface polariton (ATR, optical grating) $G_L \cdot G_S \lesssim 10^4$; /36/	"simple" image dipole effect (($E \approx 10^7$; /211/))	vibrational modulation of charge transfer to the metal (E: $10 - 10^2$; /225/)	increase of surface area by graphitic carbon overlayers (($E \approx 10^4 - 10^5$; /249/))
small particle plasmon resonances (colloids) $G_L \cdot G_S \lesssim 10^6$; /179/	image field models including renormalization of upper molecular level (($E \lesssim 10^6$; /228/))	vibrational modulation of small particle resonance (by charge injection/withdrawal) (($E \approx 10^8$; /132/))	
collective electron resonances (island films) $G_L \cdot G_S \lesssim 10^4$; /196/ $G_L \lesssim 10^2$; /197/	"advanced" image field models (($E: 1 - 10^4$; /212/))	"surface chemistry" effects: formation of complexes, radicals with new electronic properties E: ?	
collective electron resonances, optical conduction resonances ("bumpy" surface) $G_L \cdot G_S$?	vibrational modulation of plasmon surface polariton resonance (by modulation of tunneling barrier potential) ($E \lesssim 10^4$; grating /39/)		
roughness mediated, near field driven Stokes emission (statistical surface roughness) $G_S \lesssim 10^3$; /206/	vibrational modulation of metal surface polarizability via coulomb interaction ("Raman reflectivity") ($E \lesssim 10^3$; /102,223/)	resonance Raman scattering due to chemisorption induced charge transfer excitation (E: $10 - 10^2$; /236/) (E: $10 - 10^3$; /237/)	
roughness mediated excitation and scattering of plasmon surface polaritons, nonperturbive approach $G_L \cdot G_S \approx 10^2$; /205/			

"Chemical" mechanisms require contact of metal and molecule, i.e. chemisorption. They may be especially pronounced at sites of atomic scale roughness, i.e. at so-called "SERS active sites". Contrary to electromagnetic mechanisms, they are quite individual for every adsorbate/metal system. "Orchid" mechanisms might contribute to the enhancement in special situations (see, e.g.,/249/), but are certainly of limited utility for the general interpretation of SERS. The enhancement factors given in Table 1 have been estimated by the different groups for a "typical situation" (silver, green light excitation). They are set into brackets, if rather unrealistic parameters or only a crude theoretical approach have been used. Two brackets are used, if either holds. Generally, numerical estimations tend to start from highly idealized systems and therefore often yield too large values with little connection to the experimental situation. We note that many mechanisms of Table 1 may work simultaneously for appropriately prepared surfaces resulting in very large (theoretical) enhancement factors. A critical valuation of various models and numerical estimations may be found elsewhere (e.g. /67/).

2.3 Present State of Understanding

It is now generally accepted that several processes may contribute to the overall enhancement of the Raman signal from adsorbed molecules. Electromagnetic enhancement always contributes to SERS, if the surface morphology and dielectric properties of the metal allow the excitation of not too strongly damped surface plasmon resonances. *Long range* electromagnetic effects play a major role for suitably roughened surfaces /48,98/ as has been soundly demonstrated by spacer experiments /255 - 258/. There is also clear experimental evidence for an additional *short range* first layer effect /48,98,99,136/. This contribution might be particularly pronounced for or even restricted to specially adsorbed molecules, i.e. molecules on certain SERS active sites /67,87,99,259,260/. The nature of these active sites is unclear. Atomic scale roughness might be of importance /67/, at least for certain systems (e.g. for pyridine on Ag, see Chapt. 4). *Strong* SER signals are expected, if several enhancement mechanisms work simultaneously as for instance for pyridine on coldly evaporated silver films (see Chapt. 4).

Each adsorbate/adsorbent system has to be treated individually. The share of the various mechanisms contributing to the overall enhancement might be quite different for different adsorbates on the same surface or for the same adsorbate on differently prepared surfaces. The relative weak pyridine signal from silver gratings /48/ and photochemically roughened silver (/98/; Fig. 2) is presumably mainly caused by a weak long range electromagnetic effect. Coldly evaporated films, on the other hand, do not exhibit *long range* electromagnetic enhancement (Chapt. 4). The pronounced first layer effect of the strong "surface" pyridine signal from these surfaces is probably caused by a chemical and a *short range* electromagnetic effect. For low reflectivity materials like nickel and palladium, finally, any electromagnetic effect is certainly of little importance.

Electromagnetic mechanisms are in principle understood. They explain, why suitably roughened surfaces of metals of high reflectivity are the best enhancers. Qualitatively, measured SER excitation profiles (Chapt. 4), mode selective enhancement /261/, and breakdown of selection rules /116 - 118/ may be understood within a "classical" frame. The quantitative description of real, SERS active systems is in many cases, however, still marginal because of the crudeness of the models and the limited information on the surface morphology from the experiment. We emphasize that many, but not all, aspects of SERS can be understood on a purely electromagnetic basis (/67/, Chapt. 4). As so far appreciably enhanced Raman signals from LEED-clean, single crystalline, smooth surfaces have not been observed /25,28/, a major contribution of the field effects listed in Table 1 to SERS is doubtful. The chemical effect in SERS is, however, well established (see, e.g., /67/ and Chapt. 4). As it depends on the details of the metal-molecule interaction, it may be sensitive to the adsorbate and the adsorption site. Molecules bonded to certain defect sites are often subject to a particularly strong chemical enhancement (e.g. pyridine on Ag; /67/ and Chapt. 4). The details of the chemical mechanism are still a matter of debate. Currently, photon driven charge transfer excitations /262/ at sites of atomic scale roughness /263/ are thought to play a major role /67,264,265/. Qualitatively, chemical effects can account for many experimental observations (breakdown of selection rules, mode and species selective enhancement, etc. /67/; as in real systems usually chemical and electromagnetic effects contribute to SERS, it is, however, very difficult to disentangle the responsibilities of either mechanism). Quantitative theoretical evaluations are extremely complicated and represent presently hardly more than crude order of magnitude estimations. To understand the details of any chemical mechanism in SERS means to understand chemisorption, which still requires very much experimental and theoretical work.

Nevertheless, it seems worthwhile to use SERS as surface analytical tool. As long as the details of the enhancement mechanism are still unknown, extreme care has, however, to be taken when interpreting SER spectra.

3. Experimental

3.1 Arrangements

Standard optical and vacuum equipment can be used for SER studies. A typical experimental set-up is sketched in Fig. 4. Radiation from an Ar- or Kr-ion laser is
cleaned from plasma lines by means of a laser filter monochromator, polarized parallel to the plane of incidence by a polarization rotator, and focused on the sample by a cylindrical lens to a line focus of typically $0.1 \cdot 3 \ mm^2$. The angle of incidence is set to maximize excitation efficiencies ($\approx 75^{\circ}$ to the normal /24/; often
also conventional backscattering geometry, i.e. perpendicularly incident light, is
used). The power incident on the sample is typically 100 mW. The scattered light

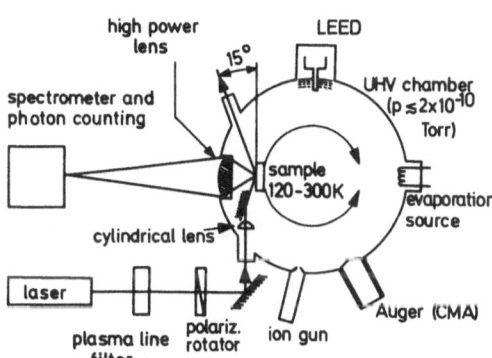

Fig. 4. Sketch of typical arrangement for SER studies in UHV. After /25/

is collected by a strong lens (f/1.0) in a cone normal to the surface and focused
on the entrance slit of the spectrometer (usually a double-monochromator with cooled
GaAs photodetector). In /27/ the scattered light is collected at $\approx 55^{\circ}$ to the surface normal to optimize also the detection efficiency /24/ (for a perpendicular
Stokes dipole). The spectrometer is set to typically several cm^{-1} bandpass, and
conventional photon counting equipment is used for signal processing. To study the
time development of SER features /115/ or very weak spectra /27/, the multiplex

advantage of a parallel detector (OMA system) has been utilized to reduce the data acquisition time.

The UHV system displayed in Fig. 4 allows to work at pressures in the low 10^{-10} Torr range. Like all surface analytical investigations, SER studies at considerably higher pressure (e.g. at $\approx 10^{-5}$ Torr /157/) suffer from possible uncontrolled adsorption of impurities. The sample holder can be cooled down close to liquid nitrogen temperature (some other set-ups permit to work at liquid helium temperature /136,266/). Metal films can be deposited onto (cooled) substrates from dc-heated Ta or W filaments. Additional facilities for surface preparation (rastering ion beam system) and characterization (Auger CM analyzer, LEED optics) may be used if required. Frequently, UPS measurements accompany UHV SER studies, and in some cases XPS, $\Delta\Phi$, TDS, and EELS have also been employed.

Freshly prepared surfaces are usually exposed to the adsorbate by back-filling the chamber or by using a beam doser directed onto the sample. In this review, exposure will be given in Langmuirs (1 L = $1\cdot10^{-6}$ Torr$\cdot$s) corrected for ion gauge sensitivity if not stated otherwise. Exposure is linked to coverage by either assuming reasonable values for the sticking probability and the density of adsorbed molecules /267/ or by evaluation of AES /98,101/, UPS /20,48,268/, or XPS/$\Delta\Phi$ data /97,269/. Coverage calibrations obtained with the various methods agree reasonably (within a factor of about two, pyridine on Ag). A direct measurement of the coverage is possible with a quartz crystal microbalance /266/.

Other, special experimental conditions and coating techniques will briefly be described when presenting corresponding SER results.

3.2 Sample Preparation and Characterization

As outlined in Sect. 2.1, a variety of differently prepared metal surfaces can display SERS. SERS active metal/gas interfaces may be divided into five classes, namely (i) coldly evaporated films, (ii) island films, (iii) grating surfaces, (iv) photochemically roughened surfaces, and (v) other (e.g. sputter-cleaned surfaces).

i) *Coldly Evaporated Films.* In 1979 the first SER signal from a surface prepared in UHV has been reported: a coldly evaporated silver film had been exposed to carbon monoxide /61/. Coldly evaporated silver films are typically thick layers ($\gtrsim$ 200 nm) deposited with $\approx$ 1 nm/s onto a polished copper substrate cooled to 120 K /99,100/. Change of the substrate to sapphire /270/, Ag /271/, or Al /136/, moderate variation of the deposition rate (by a factor of $\lesssim$ 5 /272/), as well as lowering of the substrate temperature to liquid helium values /136,142/ does not greatly affect the SERS activity. Coldly evaporated silver films give rise to strong SER signals whose intensity is similar to that from electrode surfaces. Correspondingly prepared films of Cu /123/, Au /123/, Na /142/, Li /133/, and Al /144/ are also SERS active. However, no enhanced Raman scattering could be detected from coldly evaporated palladium films /272/. The SERS activity of silver films is irreversibly lost upon annealing to room temperature /100,108,134/, and films deposited at room temperature are SERS inactive. This has been explained with the annealing of SERS "relevant" surface roughness /100,118,134/. Qualitatively, Cu behaves similarly. Here the SERS activity disappears at higher temperature compared to Ag /273/. In fact, in certain cases Cu films deposited on substrates at room temperature may exhibit enhanced Raman scattering /273/. The higher stability of Cu surface roughness /274,275/ might explain these observations.

As all investigations have to be performed at low temperature, in situ measurement of surface roughness of coldly evaporated Ag films by electron microscopic techniques is difficult (it might, however, be feasible for the more stable SERS active Cu films). In an attempt to stabilize SERS relevant roughness, silver films have been overcoated with a thin layer ($\approx$ 1 nm) of aluminum oxide and subsequently warmed to room temperature /276/. Transmission electron microscopy revealed a high density of roughness features with lateral dimensions of $\approx$ 10 - 20 nm on these surfaces, whereas an uncoated silver sample was smooth on this scale. It is, however, at present unclear, whether the Al_2O_3 coating itself contributes to the observed features. Atomic scale roughness (defect concentration) of coldly evaporated Ag films has recently been investigated with UPS /277/. A significantly larger defect concentration on SERS active samples ($\approx$ 10 - 20 % /277/) compared to films annealed to or deposited at room temperature ($\approx$ 1 % /277,278/) has been found. To gain further information on the local surface structure, UV photoelectron spectroscopy of Xe-decorated /279,280/ SERS active surfaces might be useful.

Optical properties and their relation to structural irregularities of coldly evaporated Ag films have recently been reviewed in detail /67/. With respect to annealed samples, additional absorption centered at $\approx$ 2.4 eV has been observed by several groups /133,239,281,282/. Figure 5 displays the influence of annealing /281/ and of a thin dielectric coating /283/ on the extra absorption (this quantity is approximately given by $\{1 - [R(T)/R(295 \text{ K})]\}$; /281/). The absorption maximum

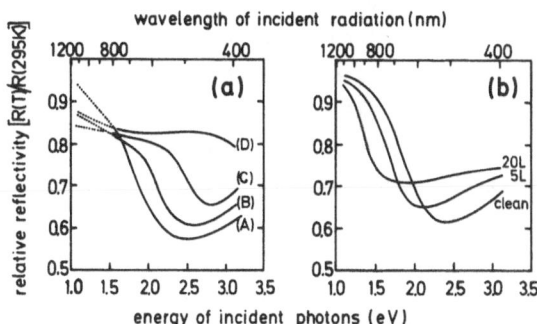

Fig. 5. Relative normal incidence reflectivity of coldly evaporated Ag films. (a) Annealing: (A) T_s = 120 K; (B) warmed up to 209 K, recooled to 120 K; (C) warmed up to 246 K, recooled to 120 K; (D) warmed up to 295 K. (b) Influence of pyridine exposure. After /281,283/

is shifted to the red by the dielectric overlayer. This indicates a surface contribution due to surface plasmon resonances, for which such shifts are expected
(see, e.g., /42,267,284/). It supports the view of /133/, where the extra absorption has been attributed to excitation of surface plasmon resonances in small,
closely packed bumps on the surface. On annealing, this structure is transformed
to a lower density of somewhat larger bumps. Simultaneously, the degree of roughness is reduced. This picture /133/ is consistent with the measured annealing of
the extra absorption (Fig. 5). On the other hand, absorption features similar to
those displayed in Fig. 5 have been ascribed to structural disorder in the bulk
(e.g. to grain boundaries /285 - 287/; opaque silver film deposited on sapphire
at 140 K). Likewise, increased resistivity of silver films deposited at low
temperature /288,289/ has been assigned to scattering of electrons by grain
boundaries, bulk point defects, and dislocations /67/ (in addition to scattering
at surface roughness). Finally, the observation of first order Raman scattering
from bulk zone boundary phonons in coldly evaporated Ag /100/, Cu, and Au /123/
(disorder induced Raman scattering, DIRS) can only be explained with bulk point
defects, where momentum conservation is relaxed. Point defects (divacancies) in Ag
have been shown to anneal at the same temperature /290 - 293/ as DIRS from Ag which
further corroborates this interpretation.

In summary, coldly evaporated films are highly disordered. A relatively great
density of defects on an atomic scale (divacancies and grain boundaries in the bulk;
adatoms at steps, kink sites, steps at the surface) is presumably accompanied by
small scale surface roughness (bumps with lateral dimensions $\lesssim$ 25 nm). Both types
of roughness features probably contribute to the measured optical properties. The
variety of disorder allows several enhancement mechanisms to work simultaneously
which should lead to strongly enhanced Raman signals (which is observed, see following chapters). It is difficult to experimentally discriminate the various contributions to the overall enhancement. An attempt to quantitatively discern the electromagnetic contribution from the chemical part is described in Chapt. 4.

ii) *Island Films*. Silver island films have been used for SER studies by several
groups /91,93,266,294/. They are prepared by slow deposition ($\approx$ 2 nm/s) of Ag onto
a glass substrate at room or slightly elevated temperature (e.g. at 390 K /266/).
Depending on the average thickness of the deposit, islands of different size can
be made /295/. With increasing mean thickness, the particles start to coalesce (at
$\approx$ 8 nm /295/) and eventually form a continuous layer (at $\approx$ 20 nm /295/). Figure 6
shows a scanning electron micrograph of a silver island film of 10 nm average thickness /266/. A variety of sizes and shapes can be seen. Uniformly sized and shaped
particles in an ordered array can be prepared by Ag evaporation onto a lithographically produced microstructure (Fig. 6, /137/).

Due to excitation of the transverse collective electron resonance /113/, silver
island films exhibit an absorption band in the visible /93,296/, which shifts from

Fig. 6. Scanning electron micrograph of silver island film (left, calibration
bare is 0.5 µm, after /266/) and of lithographically prepared microstructure
(right, calibration bare is 1 µm, after /137/)

the blue towards the red with increasing mean thickness /296/. Because of the fairly
broad distribution of sizes, shapes, and orientations of the islands, and due to
various island-island distances, the resonance is inhomogeneously broadened. This
is especially pronounced for samples consisting of large islands /296/. Depending
on the island size, the imaginary part of the silver dielectric function is usually
greater than the bulk value, which reduces the quality of the resonance further
/113/.

   Optical properties of island films are well known from the extensive theoretical
and experimental work of a Japanese group /297 - 300/. With "realistic" island film
parameters /300/, which take into account all the effects just mentioned, an electromagnetic enhancement factor due to the transverse collective electron resonance of
$G_L \cdot G_S \lesssim 10^3$ has been estimated /67/.

   Although other materials like, e.g., Cu /295/ and Au /193/, also form island
films when prepared appropriately, only few SER studies were devoted to these materials (e.g. /62,145/).

iii) *Grating Surfaces.* Several techniques have been used to prepare grating surfaces. Either, photoresist layers deposited on suitable substrates were exposed by
the interference pattern of two coherent light beams from an Ar-ion or He-Cd laser
and subsequently developed /301/. The resulting nearly sinusoidal corrugation /302/
was then coated with a thick silver layer /46/ or tunnel junction structure /51/,
or was transferred to the substrate by ion milling before coating with Ag /50/ (for
studies in UHV the photoresist has to be removed from the sample). Or, X-ray litho-
graphic techniques followed by chemical polishing have been employed to inscribe a
periodical corrugation into a Ag(111) surface /48/ (grating vector along [111] di-
rection). After several argon sputtering-annealing cycles (T ≈ 500 K) in UHV an al-

most sinusoidal modulation of the surface (still ≈ 90% (111) orientation) with a terrace width to step height ratio of ≈ 10 : 1 has been obtained /268/.

For Raman studies of adsorbed molecules, the incident beam has usually been adjusted such as to resonantly excite extended plasmon surface polaritons at the silver grating surface. Maximum coupling to the surface modes is achieved for a certain corrugation depth h, if other parameters are fixed (/42,303,304/; properties of gratings are reviewed in /305,306/). For SER studies, gratings with h/Λ between 0.06 /50/ and 0.1 /48/ have been used (Λ: grating period). No attempt has been made to optimize the coupling: best performance is expected for $h/\Lambda \approx 0.05$ (/50,303/; silver, Λ = 600 nm, λ = 514.5 nm). The Stokes emission is usually collected at a rather arbitrary angle to the grating normal. Again, no attempt has been made to systematically optimize the detection angle (see /173/). These experimental shortcomings might explain part of the discrepancy between measured and calculated enhancement factors for grating surfaces (see Chapt. 2).

iv) *Photochemically Roughened Surfaces.* Silver single crystal surfaces may be roughened in UHV by simultaneous exposure to iodine vapour ($\approx 10^{-7}$ Torr) and unfocused radiation of an Ar-ion laser (488 nm, $\approx$ 200 mW/cm^2, $\approx$ 5 min; /98,307/). Scanning electron micrographs revealed Ag particles of $\approx$ 50 nm radius separated by $\approx$ 150 - 300 nm on these surfaces /258/. As determined by AES, some iodine remained on the surface. "Electromagnetic antenna-like enhancement of the small Ag particles" /98/ is believed to be responsible for the major part of the estimated $\approx 5 \cdot 10^4$ overall enhancement (pyridine adsorption /98/), which is of long range: saturation of the SER signal occurs at $\approx$ 7 layers ($\approx$ 3.5 nm) /258/. As mentioned earlier, the estimation of the enhancement factor might be somewhat optimistic.

No optical absorption or reflection spectra have been reported for photochemically roughened surfaces. Almost nothing is known on the atomic scale roughness, i.e. the nature and concentration of possible SERS active sites (the problem is briefly touched in /258/).

v) *Other.* This class contains mechanically abraded and ion bombarded silver surfaces. Relatively weak Raman signals from predominantly the first layer of adsorbed pyridine have been obtained from polycrystalline Ag foils /97/, which were mechanically polished to mirror finish and cleaned by Ar-ion sputtering in UHV (500 V, 2 µA/cm^2, 6 hours; /97/). Scanning electron microscope photographs displayed elevated plateaus (100 - 300 nm diameter) separated by 1 - 3 µm. Between the plateaus, the surface seemed to be covered with 20 nm-diameter balls (separation ≈ 20 - 40 nm) as well as ≈ 40 nm balls (separation ≈ 100 - 500 nm). XPS spectra of the pyridine coated surface revealed a splitting of the N 1s peak, which was assigned to pyridine bonded to defect sites (unspecified) and pyridine on low-index areas /260/. The significance of the former species for SERS is outlined in Chapt. 4.

The surface topography of the sample used in /96/ is completely unknown. Here a polycrystalline silver cylinder was mechanically polished, heavily bombarded with

Ar-ions in UHV (3 keV, 10 $\mu A/cm^2$, 2 hours), cooled to liquid nitrogen temperature while still sputtering, and finally exposed immediately after sputtering (to pyridine). It is not clear, whether the observed Raman features are enhanced (see also Chapt. 4).

A procedure used in the early days of SERS, mechanical roughening (1.5 μm cerium oxide abrasive) and immersion in alkaline KCN solution, yielded a strong Raman signal from adsorbed cyanide /90/. Again, nothing is known of the morphology of this surface.

Finally, we mention the very interesting work on colloidal metal particles in solid adsorbate/argon matrices /132,308/. Spherical Ag microcrystallites (diameter 1 - 20 nm) were prepared by gas aggregation techniques /309,310/ using argon in a dynamic flow system. These can be isolated /311,312/ at liquid helium temperature in an Ar matrix (or adsorbate/argon or pure adsorbate matrix). The excitation profile of Raman lines qualitatively followed the optical extinction of these samples, which suggests electromagnetic enhancement due to surface plasmon resonances /132/. The coincidence is, however, not always observed /129/. Therefore it might be necessary to consider a non-electromagnetic mechanism of different weight in different systems as outlined in /67/.

4. Pyridine Adsorption

Pyridine has been used quite frequently in infrared transmission /313/ and laser Raman /80,314,315/ studies of the surface acidity of catalysts. Some lines are sensitive to the environment of the molecule, so that vibrational spectroscopy allows to distinguish between for instance Lewis and Brønsted acid sites, which are believed to be active sites on cracking catalysts (/316,317/; the concept of active sites in catalysis - not to be mixed up with SERS activity - is treated in several review articles /318 - 321/). The strong pyridine breathing vibration is especially suited for such investigations /315/. Therefore this molecule has also been used in the first Raman study of adsorption on electrodes /57/. The surprisingly intense spectra were strongly enhanced as was shown later /55,56/. Because of the strength of the SER signal and the sensitivity of the spectral features to the local environment, pyridine has subsequently been used in many SER studies.

Pyridine (C_5H_5N) has a planar structure of C_{2v} symmetry. Structural parameters like bond lengths and angles are known quite accurately from microwave spectroscopy /322 - 325/. The relatively large dipole moment of pyridine (2.15 D; /326,327/) has been determined from quantitative Stark effect measurements. The molecule has 27 fundamental modes of vibration. 19 of these are planar ($10A_1 + 9B_1$), 8 are non-planar ($5B_2 + 3A_2$). All vibrations are Raman active, but only 24 are infrared active (those of classes A_1, B_1, B_2). The vibrational spectra of vapour phase /328,329/, neat liquid /328 - 332/, aqueous solution /55,123,333,334/, and solid /335,336/ pyridine and deuterated pyridines have experimentally been studied in great detail. From these investigations a rather safe assignment of the fundamental frequencies and determination of the force field /337 - 339/ emerges. Only minor uncertainties remain /333/ (in this article we shall adopt the notation of /337/ for the fundamentals). With respect to adsorption of pyridine on metal surfaces, vibrational properties of pyridine-metal complexes are of particular interest. The effect of coordination on pyridine skeletal vibrations as well as on the metal-ligand stretch has been studied for a variety of complexes /340 - 347/. On complexing, pyridine vibrational frequencies generally increase, and A_1 modes are most strongly perturbed.

The electronic spectrum of the free pyridine molecule is well known /348,349/. The intramolecular excitations are only weakly perturbed and broadened upon adsorp-

tion on Ag(111) as shown recently /262,350,351/. As for the free molecule, the lowest
transition is observed at ≈ 4.1 eV in EELS spectra /349,350/ (this is an optically
forbidden singlet-triplet transition for the free molecule; the lowest allowed transition is measured at 4.9 eV). An additional, onset-like shaped, weak and broad
feature, which peaks at ≈ 2.3 eV, appeared in the spectra of adsorbed pyridine /262,
350/. It has been assigned to a metal to molecule charge transfer state analogous
to the known intramolecular charge transfer states (see, e.g., /352/). This transition is more pronounced, when pyridine is adsorbed on the irregular surface of coldly evaporated silver films /263/. Therefore it has been interpreted as charge transfer excitation localized at sites of microscopic roughness, whose density is small
on Ag(111).

Orientation and bonding of pyridine to Ag(111) have been studied by combining
UPS with EELS and TDS /20,353,354/. At ≈ 140 K and for exposures < 0.4 L, pyridine
is weakly chemisorbed through the π orbitals. Desorption of this species is observed
at ≈ 210 K. For larger exposure (> 0.5 L, 140 K), pyridine molecules are forced into
an upright orientation. This high coverage compressional phase is even more weakly
bound to the metal via the nitrogen lone-pair orbital and should therefore desorb
below 210 K. In another paper /355/, desorption of condensed multiple pyridine layers from Ag(110) is observed at ≈ 190 K, whereas a chemisorbed nitrogen-bonded species is still present at the surface in sub-monolayer amounts at 275 K. Similarly,
pyridine is adsorbed with its aromatic ring perpendicular to the surface on Cu(110)
/356/ and Ir(111) /357/ at room temperature. Little is known on the bonding of
pyridine to irregular metal surfaces such as coldly evaporated films, i.e. of bonding to defect sites. For metallic catalysts, pyridine is a relatively toxic sub-
stance /316/. Due to the nitrogen lone-electron pair it seems to block active centers by forming a relatively stable bond to these sites /314,315/.

4.1 Coldly Evaporated Silver Films

4.1.1 General Spectral Features

Figure 7 displays surface enhanced Raman spectra from coldly evaporated silver films
exposed to ≈ 0.2 L of pyridine and deuterated pyridine at 120 K /358/ (an exposure
of 0.2 L corresponds to ≈ 0.1 monolayer coverage /267/). An enhancement factor of
several 10^4 is estimated by comparing SER line intensities to corresponding values
in ordinary spectra from a thick pyridine layer condensed on a SERS inactive silver
surface at 120 K (Fig. 7c and /99/; note, that this estimation neglects the flat
metal surface contribution as well as any electromagnetic effect of the inactive
surface due to residual roughness). Different vibrations experience evidently different enhancements. This mode specific behaviour is particularly clear for the C-H

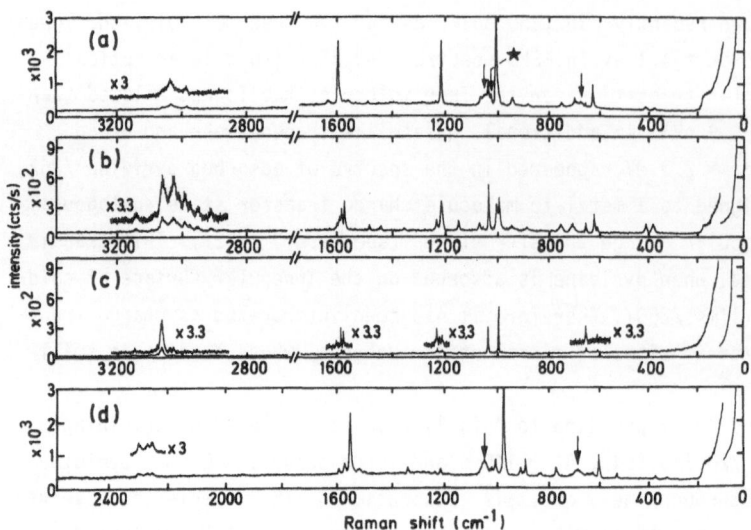

Fig. 7. Raman spectra from vapour deposited silver films. (a): coldly evaporated film exposed to 0.2 L of pyridine; (b): like (a), but exposed to 200 L; (c): film condensed at room temperature and exposed to 200 L of pyridine at 120 K; (d): like (a), but exposed to deuterated pyridine. All spectra have been taken with 200 mW of 514.5 nm radiation and 4 cm^{-1} bandpass. After /358/

stretching vibrations around 3000 cm^{-1}, which are only weakly pronounced in SER spectra from coldly evaporated silver films /108/ (adsorbed hydrocarbons behave similarly, see Chapt. 5). Upon deposition of further pyridine layers, SER line intensities *decrease* and several new peaks appear (/99/ and Fig. 7b; ≈ 10^2 layers of pyridine). Spectral positions of the new features are virtually identical to those of corresponding vibrations from a thick layer on a SERS inactive silver surface (Fig. 7c and Fig. 8). They are assigned to scattering from "bulk" pyridine, i.e. pyridine in the second and consecutive layers as well as pyridine *physisorbed* to Ag /67,119,267/. On the other hand, low coverage SER lines are due to "surface" pyridine, which are molecules *chemisorbed* to silver, probably on certain "active" sites /67,119,267/. Bulk and surface signals overlap for most vibrations, for instance for ν_6 at 1034 cm^{-1} (Fig. 8). A few modes allow, however, a separation of either contribution because of stronger chemisorption induced shifts of the vibrational energy. For example, bulk signals are observed at 993 cm^{-1} (ν_1, symmetric ring breathing) or 607 cm^{-1} (ν_3, planar ring deformation), whereas corresponding surface signals appear at 1003 cm^{-1} and 621 cm^{-1} respectively (Fig. 8b; note, that the line frequencies of surface pyridine shift somewhat with coverage, see also Sect. 4.1.2). A comparison of bulk and surface pyridine intensities reveals the pronounced "first layer effect" in SERS from coldly evaporated silver films /99,123,267/: *only surface pyridine is subject to the full enhancement of ≈ 10^4.*

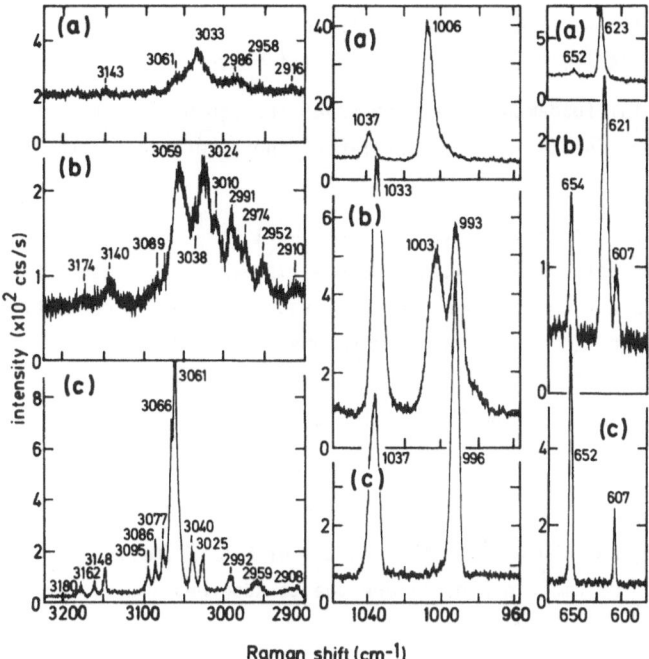

Fig. 8. Details of Raman spectra from vapour deposited silver films. (a) coldly evaporated film exposed to 0.2 L of pyridine; (b): like (a), but exposed to 200 L; (c): room temperature deposited film exposed to $2 \cdot 10^4$ L of pyridine. Same experimental conditions as for Fig. 7 (except for (c): 2 cm^{-1} bandpass). After /358/

The spectra displayed so far (Figs. 7 and 8) have been recorded by using 514.5 nm excitation. Other excitation frequencies lead to similar SER spectra from surface pyridine (Fig. 9). Note, however, that the relative line intensities depend on excitation wavelength. Modes of large vibrational energy (e.g. ν_4 and ν_5) become more prominent when changing the excitation from red to blue /123/. This points to mode

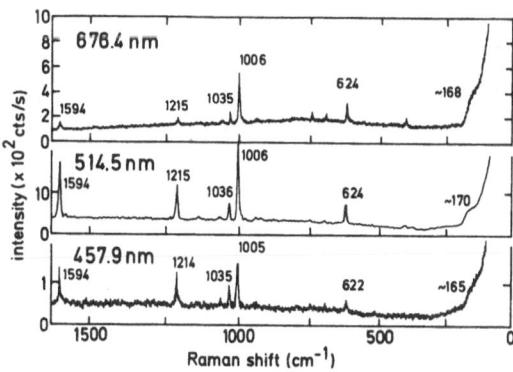

Fig. 9. SER spectra of coldly evaporated silver films exposed to 0.2 L of pyridine for different excitation wavelengths as indicated. Incident power was 60 mW (676.4 nm), 100 mW (514.5 nm), and 40 mW (457.9 nm); bandpass was set to 4.5 cm^{-1} for all spectra. After /358/

specific excitation profiles for surface enhanced Raman scattering (see Sect. 4.1.4).

A close look at the SER spectra allows the following statements:

i) Almost all pyridine skeletal fundamentals are observed and can be assigned /99, 108,119/. There is, however, no indication of a metal-pyridine stretching vibration /108/.

ii) Several low energy features at 73 cm^{-1}, 112 cm^{-1}, and 161 cm^{-1} (Fig. 10) reflect an intrinsic property of coldly evaporated silver films. They are present *without* adsorbed molecules, do not change upon pyridine exposure, and are considerably weakened upon annealing the sample to room temperature. The structures have been attributed to disorder induced Raman scattering (DIRS) from bulk acoustical phonons within the penetration depth of the light (/100,108/ and Sect. 3.2).

iii) Combination bands and overtones are weakly pronounced /358/. The first overtone of ν_1 is detected with $\lesssim$ 1% intensity of the fundamental (Fig. 11).

iv) The features at 690 cm^{-1} and 1050 cm^{-1} in the spectra of Fig. 7 (marked by an arrow) and the broad peak at 2108 cm^{-1} in Fig. 11 are due to adsorbed impurities (see Chapts. 5 - 7).

v) The line at 1026 cm^{-1}, marked by a star in Fig. 7a, is only observed after pyridine exposure, but not always (see Fig. 8a). Its intensity seems to be connected with the strength of impurity lines. The breathing vibration (ν_1) of pyridine bonded

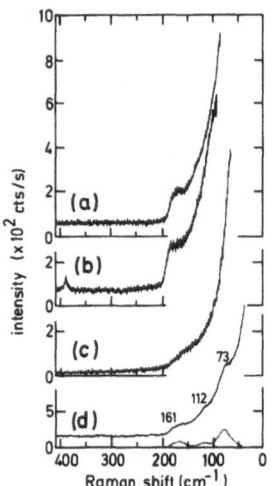

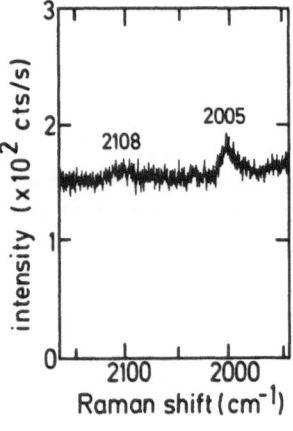

Fig. 11. Overtone of breathing mode ν_1 in SER spectrum from coldly evaporated silver film exposed to 0.2 L of pyridine (peak intensity of fundamental was 2700 cts/s). 200 mW of 514.5 nm radiation, 4 cm^{-1} bandpass. After /358/

Fig. 10. Low energy features of Raman spectra. (a): coldly evaporated film, unexposed; (b): like (a), exposed to 0.2 L of pyridine; (c): like (b), after annealing to room temperature; (d): like (a), but warmed to 220 K and recooled to 120 K to increase the intensity of the Raman features (170 mW of 514.5 nm radiation, 0.6 cm^{-1} bandpass). (a), (b), and (c) have been recorded with 200 mW of 514.5 nm radiation and 2 cm^{-1} bandpass. After /100/ and /108/

Table 2. Vibrational energies of pyridine in various systems [in cm^{-1}; number in parentheses after each vibration gives the intensity relative to breathing mode (ν_1) intensity which is set to 100 (for (g) to 10)]. (a) Neat liquid pyridine, after /331,332/; (b) 1 M aqueous solution of pyridine, after /55,123/; (c) complexed pyridine AgClO$_4$·2Py, IR study, after /341/; (d) thick layer on SERS inactive silver surface in UHV, after /358/; (e) pyridine on SERS active Ag film in UHV (0.2 L exposure), after /108,123,358/; (f) like (e), but exposed to 200 L, after /358/; (g) neat liquid pyridine-d$_5$, after /330,331/; (h) pyridine-d$_5$ on SERS active Ag film in UHV (0.2 L exposure), after /358/

Mode	(a) Liquid	(b) Aqueous Solution	(c) Complex	(d) Solid	(e) SERS on Ag 0.1ML	(f) SERS on Ag 100ML	(g) Liquid	(h) SERS on Ag 0.1ML
Symmetry	C$_5$H$_5$N	C$_5$H$_5$N	AgClO$_4$·2Py	C$_5$H$_5$N	C$_5$H$_5$N	C$_5$H$_5$N	C$_5$D$_5$N	C$_5$D$_5$N
ν_{21}, A$_2$	374 (0)			378 (1)	380 (2)	382 (30)	329 (1)	326 (1)
ν_{27}, B$_2$	405 (1)	409 (4)	412 (s)	410 (1)	413 (3)	412 (32)	371 (1)	373 (2)
ν_3, A$_1$	605 (3)	618 (15)	641 (m)	607 (3)	623 (20)	621 (68) / 607 (16)	582 (3)	601 (18)
ν_{12}, B$_1$	652 (6)	654 (30)	651 (w)	652 (10)	652 (2)	654 (34)	625 (6)	626 (2)
ν_{26}, B$_2$	700 (0)		697/700 (s)	706 (0)	696 (3)	707 (28)	530 (1)	530 (4)
ν_{23}, B$_2$	749 (0)	756 (3)	749/754 (s)	755 (0)	749 (4)	753 (28)	567	557 (0)
ν_{25}, B$_2$	886 (1)	890 (3)	889 (w)	897 (1)	880 (1)	887 (11)	762 (4)	772 (8)
ν_{20}, A$_2$	(886)						690 (5)	696 (0)
ν_{24}, B$_2$	942 (0)	950 (4)	944 (w)	958 (0)	942 (4)	944 (12)	823 (5)	
ν_{22}, A$_2$	981 (4)	980 (4)	990 (vw)	986 (2)	972 (2)	(965 (4))	798 (0)	812 (1)
ν_1, A$_1$	992 (100)	1004 (100)	1005/1012 (m)	996 (100)	1006 (100)	1003 (100) / 993 (115)	962 (10)	975 (100)
ν_6, A$_1$	1030 (74)	1037 (87)	1037 (s)	1037 (46)	1037 (20)	1033 (170)	1006 (7)	1007 (10)
ν_8, A$_1$	1068 (1)	1071 (22)	1068 (s)	1059 (1)	1069 (6)	1057 (7)	823	829 (2)
ν_{17}, B$_1$	1085 (0)			1067 (1)		1070 (28)	833 (5)	839 (2)
ν_{16}, B$_1$	1148 (1)	1154 (17)	1156 (w)	1150 (3)	1150 (3)	1149 (39)	(887)	
ν_5, A$_1$	1218 (6)	1221 (41)	1218 (vw) / 1224 (m)	1227 (10)	1215 (59)	1216 (116)	886 (5)	889 (14)
ν_{11}, B$_1$	(1218)		1233 (vw)	1217 (2)			908 (5)	909 (6)
ν_{15}, B$_1$	1375 (0)	1362 (1)	1361 (vw)	1360 (0)	1355 (1)	1356 (8)	1322	1324 (1)
ν_{18}, B$_1$	1439 (0)	1447 (4)	1440/1449 (s)		1442 (1)	1444 (10)	1301 (0)	1306 (1)
ν_9, A$_1$	1482 (2)	1491 (19)	1482 (m)	1488 (1)	1480 (2)	1481 (16)	1340 (1)	1341 (9)
ν_{14}, B$_1$	1572 (4)	1579 (33)	1573 (w)	1576 (10)	1572 (4)	1573 (103)	1542 (4)	1575 (7)
ν_4, A$_1$	1583 (6)	1597 (38)	1597/1607 (m,s)	1586 (9)	1593 (59)	1591 (44) / 1583 (74)	1530 (6)	1555 (51)
ν_{10}, A$_1$	3036 (2)			3037 (1)	3033 (4)	3038 (~ 20)	2254 (6)	2250 (2)
ν_{13}, B$_1$	(3036)			3025 (2)		3024 (35)	2285	2285 (0)
ν_2, A$_1$	3054 (26)	3076 (?)		3061 (17)	3061 (1)	3059 (40)	2293 (10)	2290 (2)
ν_7, A$_1$	(3054)			3066 (9)			2270 (2)	2266 (2)
ν_{19}, B$_1$	3083 (2)			3077 (1)		3089 (~ 3)	(2293)	

to impurity sites is presumably responsible for this peak (see also Sect. 4.1.3). The line at 1026 cm^{-1} is quite strong in SER spectra from pyridine on activated silver electrodes for potentials positive ($\approx$ 1 V) to the point of zero charge /57, 247/. Here it has tentatively been assigned to Lewis-coordinated pyridine /57/ or adsorbed pyridinium cations PyH^{+} /246/.

Table 2 summarizes the pyridine SER line energies and intensities. The mode selective enhancement is clearly seen when comparing A_1 mode intensities of surface pyridine (column e) and neat pyridine (column a). Vibrational energies of surface pyridine are generally shifted to higher values with respect to neat pyridine. The shift is most pronounced for some planar ring modes (ν_3, ν_1, ν_4), but does not exceed 20 cm^{-1}. This points to weak perturbation of the adsorbed molecule, i.e. weak chemisorption /20/. Similar line shifts are observed when pyridine is coordinated to Ag in metal-pyridine complexes (/341/ and column c in Table 2), or bonded to Ag$_x$- or Cu$_x$-clusters ($x \leqslant 3$) in an argon matrix /359/. They have been explained in terms of coupling with low frequency vibrations, particularly with the metal-pyridine stretching /339,347/, or, alternatively, with changes of the electron distribution in the molecule resulting in stronger chemical bonds in the ring system /334/. For our purposes, the correct interpretation of the shifts of complexed pyridine is less important than the fact that they exist. The similarities of vibrational features of surface pyridine and metal-pyridine complexes may allow to speculate on the adsorption geometry of the former. Bonding to surface sites of "certain acidity" via the nitrogen lone-pair orbital should be involved, an orientation as proposed in /20/ (low coverage phase) seems reasonable. The question of orientation and bonding of

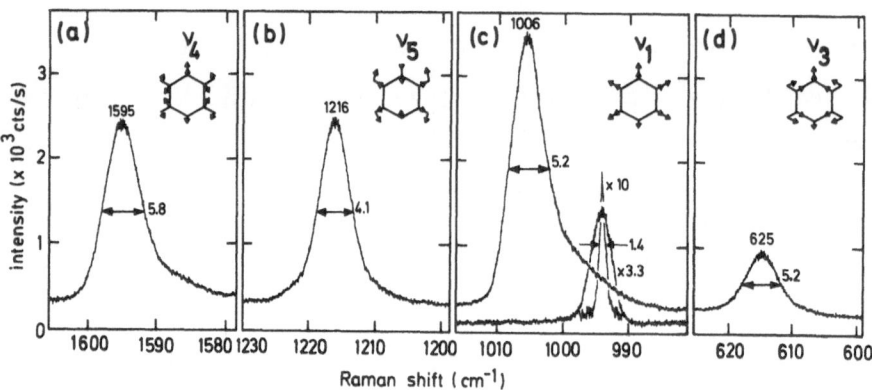

Fig. 12. Line shape of some SER lines from coldly evaporated silver films exposed to 0.2 L of pyridine. Additional peaks in (c) are ordinary lines from room temperature deposited films exposed at 120 K to 10^3 L (3.3times enlarged) and 10^5 L (10times enlarged, 0.5 cm^{-1} bandpass). Spectra have been taken with 200 mW of 514.5 nm radiation and 4 cm^{-1} bandpass. Indicated halfwidth (FWHM) has been corrected for spectrometer response. After /358/

surface pyridine and hence of the character of the SERS active species is addressed in somewhat more detail below.

SER lines of surface pyridine display quite distinct shapes /358/ (Fig. 12): ν_3 and ν_5 are almost symmetrical, whereas ν_1 and ν_4 show a pronounced asymmetry (note, that the breathing mode ν_1 in ordinary spectra is symmetrical, as expected; due to convolution by the spectrometer function measured peaks resemble Gaussian rather than Lorentzian profile, Fig. 12c). SER lines from adsorbates on coldly evaporated films are frequently asymmetrical. In general, a slow increase on the low energy side of the line is accompanied by a steeper decrease on the high energy side. An interpretation of the line shape will be given in Chapt. 5, where ethylene adsorption is discussed.

4.1.2 Coverage Dependence

Pyridine exposures as low as 10^{-2} L corresponding to roughly one per cent of a monolayer coverage result in easily detectable Raman signals from coldly evaporated Ag films /99/. Line intensity, shape, and spectral position vary with coverage, and new lines develop. The latter has briefly been touched in the preceeding section (see Fig. 7). The change of spectral features with exposure is shown in detail for the breathing mode region in Fig. 13. The bulk pyridine ν_1 line emerges from the slope of the surface pyridine line at $\approx$ 2 L exposure and is detected as a distinct peak for $\gtrsim$ 5 L. It is much stronger than expected for ordinary Raman scattering, much weaker, however, than the surface pyridine signal for 0.2 L exposure. Its intensity does not measurably increase for exposures between $\approx$ 5 L and $\approx$ 30 L (in fact, it decreases slightly). It starts to grow for exposures $\gtrsim$ 30 L caused by ordinary scattering, analogous to pyridine condensed on inactive silver surfaces /25/. Hence bulk pyridine signals at 993 cm^{-1} are weakly enhanced, and the effect is restricted to molecules in the *immediate vicinity* of the silver surface.

The development with exposure of the breathing mode intensities depends on the excitation wavelength (Fig. 14). Even for $2 \cdot 10^4$ L exposure, bulk signals do not exceed surface pyridine signals for red excitation, and the intensity of ν_6 (overlapping bulk and surface signal) is always smaller than that of ν_1. Blue and green excitation leads to features similar to ordinary spectra from thick layers on SERS inactive surfaces /25/. Here ν_6 is the strongest mode for intermediate exposures (Fig. 14, 200 L). Recalling the ω_s^4 dependence of ordinary scattering and the SERS excitation profile of the breathing modes with its peak in the red (see Sect. 4.1.4), the interpretation is straightforward: relatively small ordinary and large SER signals combine for 676.4 nm excitation, whereas the opposite is the case for blue (green) excitation.

The SER line of the breathing mode from surface pyridine broadens and shifts to slightly smaller energy with coverage (Fig. 13). Other vibrational modes behave

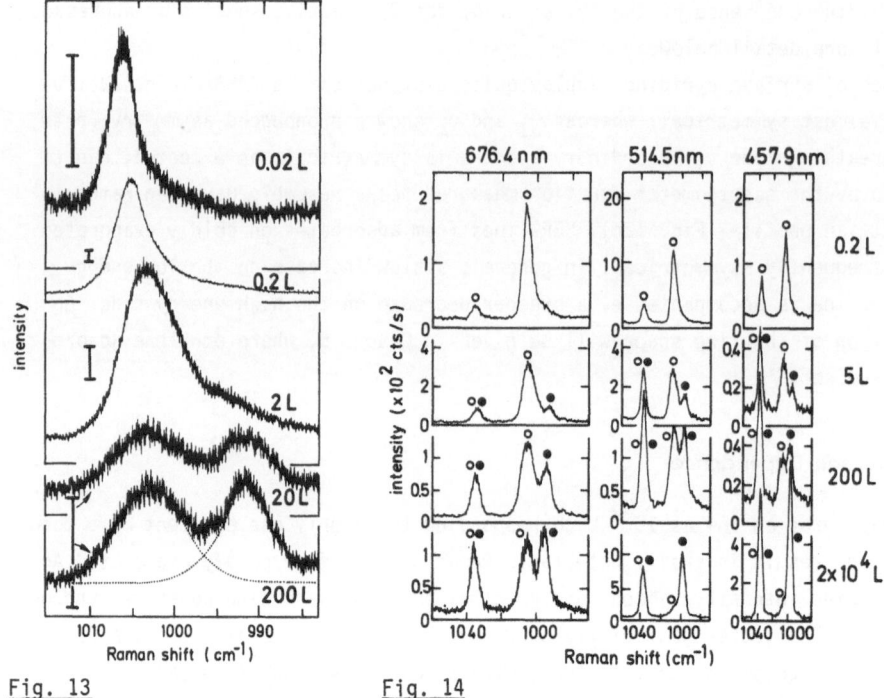

Fig. 13

Fig. 14

Fig. 13. SER spectra of symmetric breathing vibration from coldly evaporated Ag films exposed to various amounts of pyridine as indicated. 200 mW of 514.5 nm radiation, 2 cm^{-1} bandpass. The bare to the left of each spectrum represents 100 cts/s. After /358/

Fig. 14. Development of breathing modes with exposure for pyridine on coldly evaporated Ag film. Spectra have been taken with 65 mW (676.4 nm), 200 mW (514.5 nm), and 75 mW (457.9 nm). Bandpass was 4 cm^{-1} for all spectra. Circles: surface pyridine, dots: bulk pyridine. After /358/

similarly (Fig. 15). The variations of line width and spectral position are most prominent for exposures, which correspond to roughly monolayer completion. Besides these changes, one observes a transformation from the characteristic, asymmetrical SER line shape of ν_1 (0.2 L) into a more symmetrical line with exposure (200 L; Fig. 13). A detailed interpretation of these observations is difficult, since little is known of pyridine adsorption on the irregular surface of SERS active coldly evaporated silver films. As the variations are most pronounced when completing a monolayer, interaction with adjacent adsorbed molecules seems to be involved. As mentioned, the discussion of the variation of the line shape is postponed to the ethylene/Ag system (Chapt. 5).

A linear increase of the SER intensity is observed for very small exposures ≲ 0.1 L (Fig. 16). When this increase is extrapolated to greater coverage and when

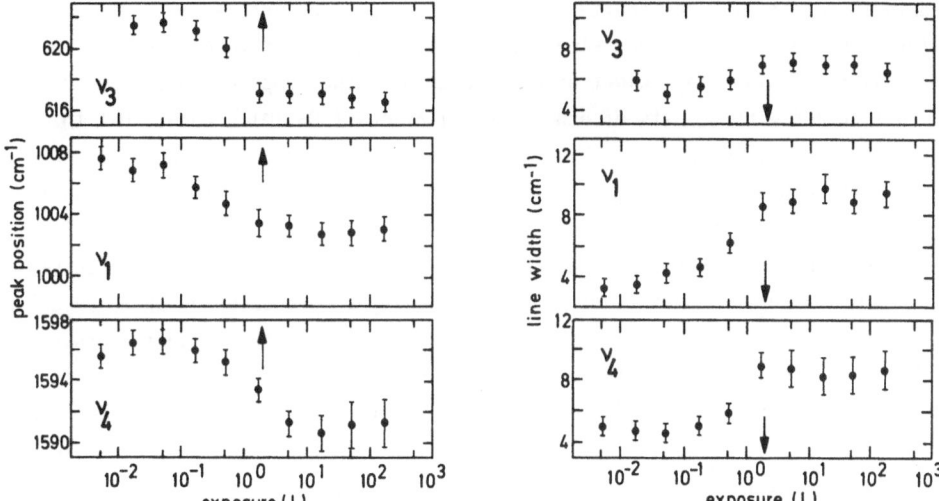

Fig. 15

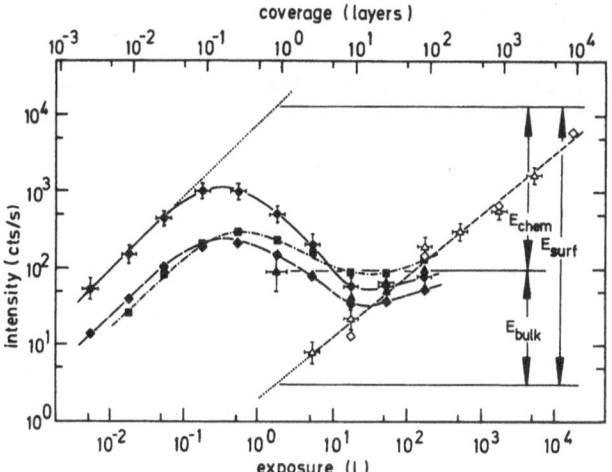

Fig. 16

Fig. 15. Shift of spectral position (left) and broadening of line width (right) as a function of exposure for three SER lines from surface pyridine on coldly evaporated silver films. Arrow marks exposure equivalent to monolayer formation. After /358/

Fig. 16. Peak intensity of some pyridine Raman lines as a function of exposure. Filled-in symbols: pyridine on coldly evaporated film /99/ [dots: ν_1; rhombs: ν_3; squares: ν_6 (all from surface pyridine); triangles: ν_1 (bulk pyridine)]. Open symbols: ν_1 of pyridine on SERS inactive Ag surface /25/ (rhombs: (110) single crystal; triangles: polycrystalline slug). Lines are guides to the eye. For an explanation of E_{chem}, E_{bulk}, and E_{surf} see text

the data for ordinary Raman scattering from SERS inactive surfaces are extended to smaller coverage, the different slopes of the two lines point to a roughly 30 per cent smaller sticking coefficient of multilayer pyridine compared to surface pyridine. This agrees with results of other investigations (/98,101/; note, that the upper scale in Fig. 16 neglects this difference). Taking the difference into account a total enhancement E_{surf} of $\approx 10^4$ for the symmetric breathing vibration of surface pyridine is estimated, which is slightly smaller than the less accurate value given above. Saturation of the SER signal from surface pyridine is observed for ≈ 0.3 L exposure. Upon further exposure, the signal decreases by up to roughly a factor of 18 for ≈ 30 L, before the intensity starts to increase again. Other surface pyridine lines, for example ν_3, exhibit a similar coverage dependence (Fig. 16). Note, however, that the intensity decrease for ν_3 after saturation is smaller than for ν_1. This is due to the fact that peak intensities rather than integrated intensities are plotted in Fig. 16. The line shape of ν_3 broadens less than that of ν_1 with exposure (see Fig. 15). Taking integrated values, either line intensity drops by roughly an order of magnitude after saturation. A slightly different exposure dependence is observed for overlapping bulk and surface pyridine signals. The antisymmetric breathing mode ν_6 shows maximum intensity at a larger dose (0.6 L), and subsequently decreases by only a factor of ≈ 4 (Fig. 16). Both effects are caused by the bulk pyridine contribution to the overall intensity of ν_6, which does not vary appreciably between ≈ 2 L and ≈ 10 L as outlined above for ν_1 (see also Fig. 16; the observed slight intensity decrease is interpreted in Sect. 4.1.4).

The short range enhancement E_{bulk} of the bulk pyridine signal is estimated to > 30 for the symmetric breathing vibration (Fig. 16). The value represents a lower limit, since the bulk ν_1 intensity used for the estimation is certainly from less than a monolayer of adsorbed molecules. Assuming that this mechanism also amplifies the surface pyridine signal, an additional effect must be responsible for the $\lesssim 300$ times stronger enhancement for this species. For reasons, which will become clear later, this factor is called E_{chem} in Fig. 16.

Finally we note, that the linear exposure dependence of the ν_1 intensity from pyridine on Ag(110) or on SERS inactive polycrystalline foils down to ≈ 3 layers coverage (Fig. 16; /25/) excludes short range, "smooth surface" enhancements of $\gtrsim 3$. Very recent measurements showed the linear development of the intensity also in the sub-monolayer region /28/, which leaves no space for *any* "smooth surface" enhancement of $\lesssim 1$ nm range, i.e. for any measurable image field effect.

36

4.1.3 Annealing Behaviour

Annealing to room temperature irreversibly destroys the enhancement properties of coldly evaporated silver films /99,100,134/. The temperature variation of the background intensity, of the Rayleigh scattered light, and of the symmetric breathing mode of pyridine has been discussed in detail in /100,239/. In these experiments the sample was warmed to room temperature with $\approx$ 1 K/min. Figures 17 and 18 summarize the annealing behaviour of Raman signals from adsorbed pyridine. The peak intensity ν_1 of surface pyridine (0.2 L exposure) first increases with temperature, exhibits a maximum at $\approx$ 210 K, and then decreases. The line disappears at $\approx$ 270 K. It cannot be restored by recooling to 120 K and re-exposing to pyridine /100/. The

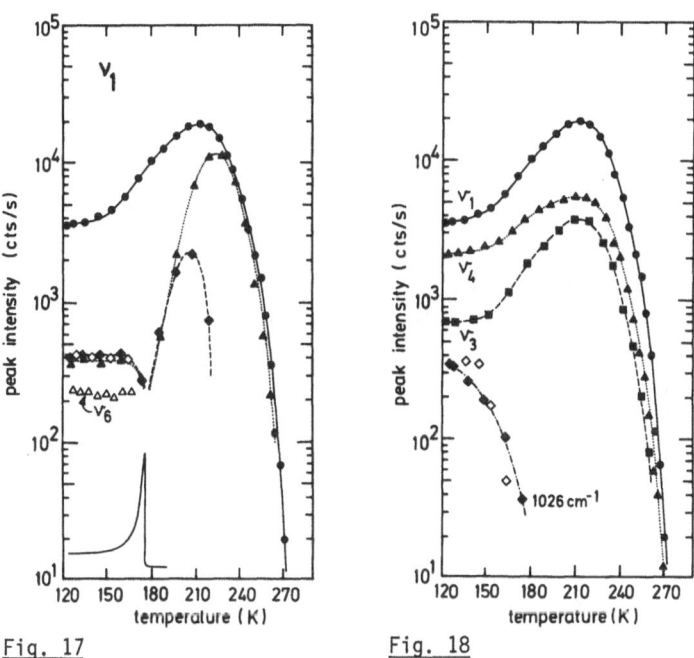

<u>Fig. 17</u>   <u>Fig. 18</u>

<u>Fig. 17.</u> Annealing of Raman peak intensities from breathing mode ν_1 of pyridine on Ag. Filled-in symbols: coldly evaporated film (dots: 0.2 L, surface pyridine; triangles: 200 L, surface pyridine; rhombs: 200 L, bulk pyridine). Open symbols: SERS inactive, room temperature deposited film exposed to 200 L (rhombs: ν_1; triangles: ν_6). Lower left inset shows temperature variation of the Rayleigh scattered light from the SERS inactive sample. 200 mW of 514.5 nm radiation, 4 cm^{-1} bandpass, and $\approx$ 1 K/min temperature variation. Lines are guides to the eye. After /358/

<u>Fig. 18.</u> Temperature variation of various SER peak intensities from coldly evaporated Ag films exposed to 0.2 L of pyridine (filled-in symbols; lines are guides to the eye). Open rhombs: measured annealing of "impurity" line at 1050 cm^{-1}. 200 mW of 514.5 nm radiation, 4 cm^{-1} bandpass, and $\approx$ 1 K/min temperature variation. After /358/

37

coldly evaporated silver film has lost its SERS activity irreversibly. As shown in /100/ and in agreement with TDS studies /355/, the decrease of intensity for $T \gtrsim$ 210 K is not due to desorption of pyridine.

When the SERS active silver film is coated with a thick pyridine layer, neither bulk nor surface pyridine Raman signals of ν_1 change between 120 K and 175 K (Fig. 17). The solid pyridine overlayer apparently prevents any annealing. Multilayer pyridine, i.e. layers beyond the first, desorb at 175 K. This follows from the disappearance of the bulk pyridine signal from *inactive* Ag surfaces, the behaviour of the Rayleigh scattered intensity (lower left inset in Fig. 17), and the pressure increase in the vacuum chamber at this temperature. Above 175 K, surface pyridine signals from the thickly coated active sample grow much faster than those from the sample exposed to only 0.2 L. They eventually approach the latter at $\approx$ 240 K. The bulk pyridine intensity from physisorbed molecules in direct contact with silver increases also for $T \gtrsim$ 180 K, but peaks already at $\approx$ 205 K, and is finally lost at $\approx$ 220 K. This is presumably due to desorption of the species responsible for the bulk pyridine line for $T \gtrsim$ 180 K. Desorption of a weakly bonded species has been observed in this temperature range /20,97/.

Intensities of various surface pyridine lines anneal similarly as shown for ν_1 and the ring deformation modes ν_3 and ν_4 in Fig. 18 (0.2 L exposure). Note, however, that the increase between 120 K and 210 K is weaker for ν_4 (factor of 3) than for ν_3 and ν_1 (factor of 5 - 6). Only the line at 1026 cm^{-1} behaves differently. Its intensity drops immediately upon warming from 120 K and it disappears at $\approx$ 180 K. Some impurity lines display a quite similar intensity variation (e.g. the line at 1050 cm^{-1} in Fig. 7; open rhombs in Fig. 18). This corroborates the tentative assignment of the line at 1026 cm^{-1} to ν_1 of pyridine bonded to impurity sites (Sect. 4.1.1).

Spectral position and halfwidth of SER lines may also vary upon annealing. Variations are insignificant for samples exposed to 0.2 L of pyridine. A line width independent of temperature and a very small line shift to larger vibrational energy have been observed (Fig. 19). This justifies the use of peak intensities in Figs. 17 and 18. It does, however, not hold for coldly evaporated silver films exposed to 200 L of pyridine. After desorption of multilayer pyridine at 175 K, a considerable decrease of the line width and appreciable blue shift of the breathing vibration of surface pyridine is observed (Fig. 19). Bands of other modes change similarly. It is remarkable, that line parameters vary in the same way with coverage, where high coverage data correspond to those at $\approx$ 180 K and low coverage data to those at $\approx$ 250 K (compare Figs. 19 and 15). This suggests a common explanation which will be discussed later. In closing we note, that the different line widths of the breathing vibration of surface pyridine from thick and thin overlayers can explain only part of the corresponding intensity difference at 120 K (Fig. 17).

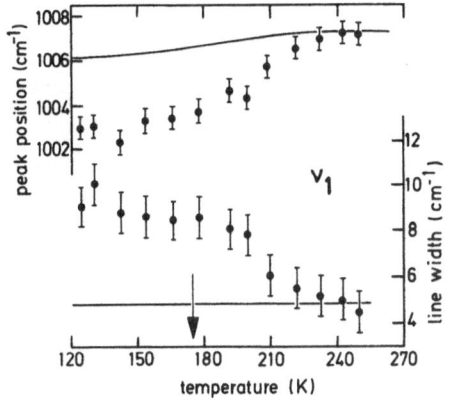

Fig. 19. Variation of spectral position and line width of ν_1 from surface pyridine with temperature. Solid lines: from sample exposed to 0.2 L (measured); dots: from sample exposed to 200 L. Arrow indicates desorption temperature of multilayer pyridine. 200 mW of 514.5 nm radiation, 4 cm^{-1} bandpass, and $\approx$ 1 K/min temperature variation. After /358/

The observed effects may reflect a complicated simultaneous acting of several processes. Surface and bulk defects as well as small scale surface roughness in highly disordered coldly evaporated silver films anneal with temperature (in general not simultaneously; Sect. 3.2). This affects SER intensities via the density of possible SERS active sites and the quality of electromagnetic resonances. In addition, spectral position and strength of the latter depend on overlayer thickness. Finally, geometry and density of adsorbed molecules may vary with temperature by, for instance, desorption of weakly bonded species, i.e. of physisorbed molecules. This may affect vibrational interaction of adsorbed molecules as well as the density of the "relevant" species, i.e. of surface pyridine. The contributions of the various processes to the intensity variation of SER signals with temperature are discussed in Sects. 4.1.4 and 4.4.

4.1.4 Excitation Spectra

Raman excitation spectra are particularly useful to elaborate various contributions to SERS. They may provide information on the intermediate electronic states of the scattering process as well as on the surface topography of the metal via the local field strength which affects the Raman scattered intensity. A variation of appropriate experimental parameters may allow to discriminate contributions of different processes to the excitation profile. Hence the share of different enhancement mechanisms might be estimated from such investigations.

The procedure employed to obtain excitation profiles has been described in detail in /119/. Thick pyridine layers condensed on SERS inactive silver surfaces served as standard. An exposure of 1.9 L of pyridine was assumed to form a monolayer of 0.5 nm thickness (for details see /267/).

Raman excitation spectra from coldly evaporated silver films exposed to pyridine exhibit resonance-like profiles (Fig. 20, /119/). For the breathing vibration ν_1, the broad resonance peaks at $\approx$ 2.15 eV (FWHM: $\approx$0.5 eV). An intensity ratio on and off resonance of $\lesssim$ 100 is estimated from the shape of the profile. Similar resonances are observed for other pyridine lines as well as for surface enhanced Raman lines of other adsorbates (Fig. 20; /119,281,360/). Independent of the adsorbed species, the observed maxima shift to shorter wavelength with increasing vibrational energy. This is summarized in Fig. 21 for different lines of various adsorbed mole-

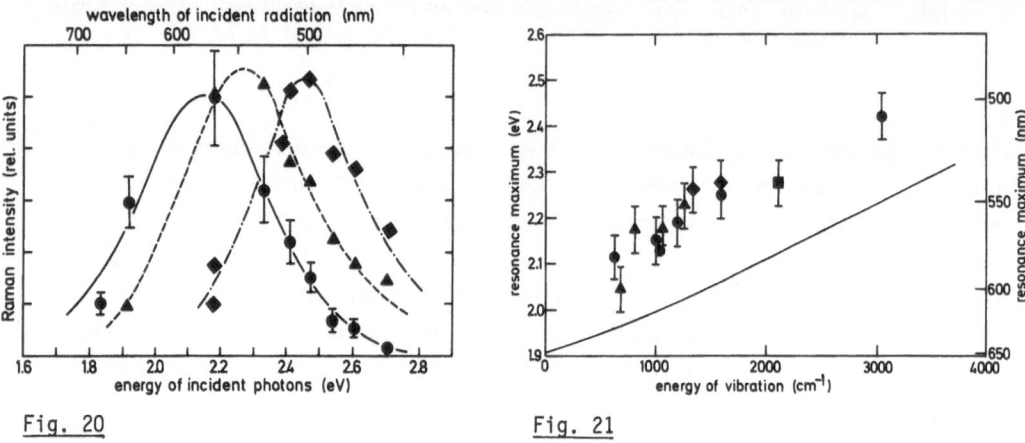

Fig. 20   Fig. 21

Fig. 20. SER excitation profiles from molecules on coldly evaporated silver films. Symmetric breathing (dots) and C-H stretching mode (rhombs) of surface pyridine (0.2 L), and symmetric scissors mode (triangles) of ethylene (36 L). Lines are guides to the eye. After /119,281/

Fig. 21. Spectral position of resonance maximum as a function of vibrational energy for various adsorbates on silver films. Dots: pyridine lines (0.2 L); triangles: oxygen lines (340 L); rhombs: ethylene lines (36 L); square: "carbon monoxide" line $(1.8 \cdot 10^4$ L; see Chapt. 6). Curve has been calculated (see text). After /281/

cules. Here the resonance maximum has been plotted against vibrational energy (the data are from excitation profiles plotted against the energy of the *incident* photons; when the resonances are plotted as a function of the *Stokes* energy, the spectral position of the resonance maximum is almost independent of the vibration /281/). Note, that the intensities of different lines for given wavelength of the incident radiation as displayed in Fig. 20 cannot be compared, since the curves have been normalized to give the same intensity for 568.2 nm excitation (intensity of C-H stretching mode attenuated by a factor of five).

SER intensities increase continuously with increasing wavelength of the incident radiation for pyridine on Cu and Au (Fig. 22; /123/). The data suggest a threshold

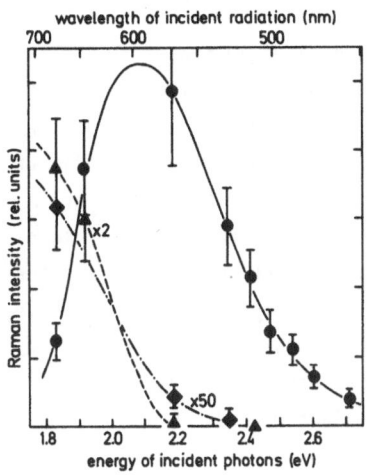

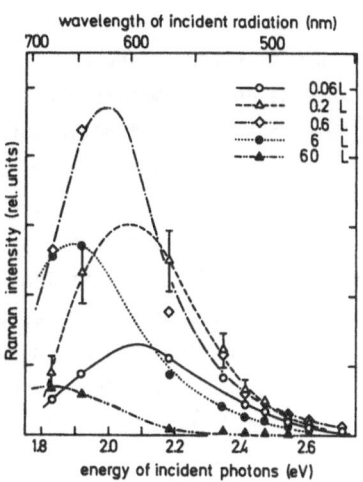

Fig. 22. SER excitation profiles for ν_1 of pyridine on Ag (dots, 0.2 L), on Cu (triangles, 2 L), and on Au (rhombs, 2 L). Note enlargement of Cu and Au data. Lines are guides to the eye. After /123,281/

Fig. 23. SER excitation profiles from surface pyridine on Ag for various exposures as indicated (symmetric breathing vibration). Lines are guides to the eye. After /267/

for SERS from these materials at $\approx$ 2.4 eV and an excitation profile maximum outside the accessible wavelength range (> 700 nm). Qualitatively, the results are similar to those from electrode surfaces where gold also exhibits relatively small intensities /122/.

Increasing exposure leads to characteristic changes of the excitation profiles of surface pyridine (Fig. 23; all results displayed in this section are from sur-

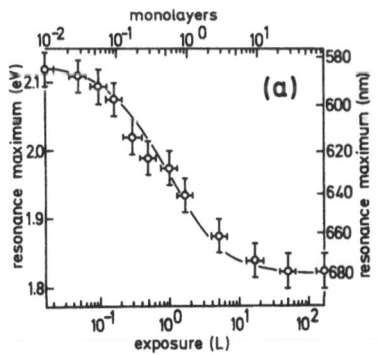

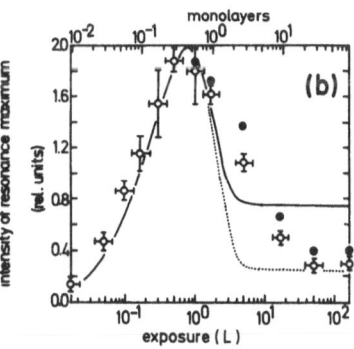

Fig. 24. Spectral position (a) and intensity (b) of the resonance maximum in the SER excitation profiles as a function of exposure (ν_1 of surface pyridine). Filled-in data points in (b) are from evaluation of integrated intensities instead of peak intensities. Curves have been calculated (see text). After /267/

face pyridine if not otherwise stated). The resonance shifts from ≈ 590 nm (2.12 eV, 0.06 L) to ≈ 680 nm (1.82 eV, 60 L). The intensity at maximum increases for small coverage, exhibits an extremum at 0.6 L exposure, and drops again. The spectral position of the resonance as a function of exposure is displayed in Fig. 24a (as in Fig. 16, the upper scale - thickness of pyridine overlayer - neglects the difference of the sticking coefficient of multilayer and surface pyridine). Note, that only ≈ 6 L exposure (≈ 3 layers corresponding to 1.5 nm thickness) are sufficient to displace the resonance by ≈ 90% of its final shift for very thick coatings. The intensity at maximum of the resonance ceases to change considerably for exposures above 20 L (Fig. 24b). The maximum of this quantity is observed for ≈ 1 L corresponding to about half a monolayer pyridine coverage. For thick overlayers, the relative intensity of equivalent surface and bulk pyridine vibrations depends on the excitation wavelength. The ratio $I_{bulk}/I_{surface}$ for three vibrational modes is displayed in Fig. 25. Whereas this quantity is almost independent of excitation wavelength for ν_3, it increases with excitation energy for ν_1 and exhibits a maximum around 520 nm for ν_4.

The influence of annealing on SER excitation profiles of the pyridine breathing vibration (ν_1, 0.2 L) is shown in Fig. 26. The resonance shifts to shorter wave-

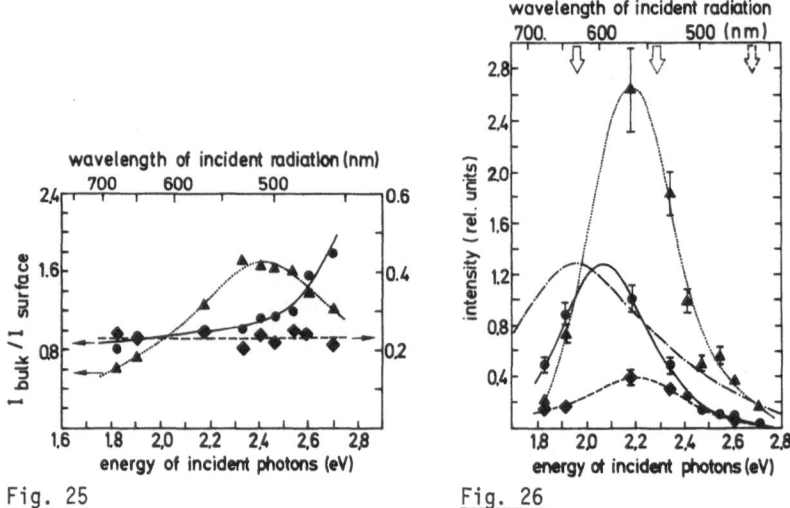

Fig. 25   Fig. 26

Fig. 25. Intensity ratio of corresponding bulk and surface pyridine lines as a function of the energy of the incident photons (SERS active film; 200 L exposure). Rhombs: ν_3; dots: ν_1; triangles: ν_4. Lines are guides to the eye. After /119/

Fig. 26. Annealing of SER excitation profiles from pyridine on coldly evaporated silver film (symmetric breathing vibration, 0.2 L). Dots, full curve: T_S = 120 K; triangles, dotted curve: warmed up to 209 K and recooled to 120 K; rhombs, dashed curve: warmed up to 246 K, recooled to 120 K. Dashed-dotted curve and arrows (position of maximum) have been calculated from absorption spectra (see text). After /281/

length with increasing temperature. The intensity at maximum first increases (up to ≈ 210 K) and then drops rapidly. In passing we note, that part of the excitation profile studies have recently been repeated /360/: the results essentially agree with those displayed here.

The observations can be explained by assuming a mainly electromagnetic origin of the excitation profile resonances. In this picture, excitation of surface plasmon type resonances in "appropriate" roughness features of the surface (bumps) leads to enhanced Raman scattering /190,191,361,362/ (see also Chapts. 2 and 3). It is assumed that "cold" evaporation creates the "appropriate" surface roughness, i.e. bumps of suitable shape and size. As this is a property of the metal, similar excitation profiles are expected for different adsorbates (see Fig. 20; the shift of the resonance is explained below). Because of the optical properties of Cu and Au /363/, electromagnetic resonances in these metals are strongly damped above ≈ 2.5 eV, which explains the results of Fig. 22, especially the threshold behaviour. Increasing the refractive index of the medium adjacent to the rough surface causes a red shift of the electromagnetic resonance (see, e.g., /364/), which is reflected by the excitation spectra displayed in Fig. 23 (see also Fig. 24). The shift contributes a factor of ≈ 3 to the 18fold decrease of the ν_1 peak intensity with coverage for exposures ≳ 0.2 L (Fig. 16, excitation wavelength 514.5 nm; according to the results of Fig. 23 the shape of the intensity versus exposure curve should change with excitation wavelength, which is indeed observed /358/). The blue shift of the excitation profile resonance upon annealing (Fig. 27) is explained with a transformation of a high density of small bumps on the coldly evaporated films to a lower density of somewhat larger bumps /118,281/.

If the rough surface is modeled in a simplifying, crude approach by an ensemble of isolated, non-interacting spheroids, quantitative comparison of some experimental results with theoretical predictions is possible.

Firstly, the spectral dependence of SER intensities can be related to the optical absorption $A(\omega)$ and the dielectric function $\underline{\varepsilon} = \varepsilon_1 + i\varepsilon_2$ of the metal by /93/:

$$I_{SERS} \sim \frac{\varepsilon_1^2(\omega_L) \cdot A(\omega_L)}{\omega_L \cdot \varepsilon_2(\omega_L)} \cdot \frac{\varepsilon_1^2(\omega_S) \cdot A(\omega_S)}{\omega_S \cdot \varepsilon_2(\omega_S)} \tag{5}$$

where ω_L and ω_S are the frequencies of the incident and Stokes photons. $A(\omega)$ may be extracted from reflectivity measurements (/239,281/; see also Chapt. 3). The relative reflectivity $[1 - R(T)/R(295\ K)]$ is a measure of the *additional* optical absorption of coldly evaporated films with respect to annealed films (Rayleigh scattering neglected, for details see /239/). $A(\omega)$ is approximated by this quantity, and excitation profiles for different vibrational energies are calculated with the help of (5). The maxima of the calculated profiles follow the solid line in Fig. 21, which nicely reproduces the experimentally observed trend. Within the used approach, the shift of the excitation profile resonance with vibrational energy is the consequence of a rather broad absorption profile of coldly evaporated films and the rapid variation of the silver dielectric function in the frequency region of interest (a detailed discussion is given in /281/). Similarly, the annealing behaviour of excitation profiles (Fig. 26) may be calculated from corresponding relative reflectivity spectra (Fig. 5 and /281/) by using (5). The dashed-dotted line in Fig. 26 is the calculated spectral dependence of I_{SERS} (no parameters except the height of the curve have been fitted). We find reasonable agreement between theory and experiment. Quantitative agreement between calculated and measured shift of the excitation profile maxima with annealing temperature is, however, poor (arrows in Fig. 26 mark the calculated peaks in I_{SERS}). In addition, relative reflectivity spectra (Fig. 5) do not show an equivalent to the increase of the SER intensity on annealing to 209 K. This might be explained with partial masking of the "SERS relevant absorption" in the reflectivity spectra by other absorption processes /281/ and/or the influence of effects not considered so far, e.g. a temperature dependent density of SERS active molecules (sites) /267/ on the silver surface /100/.

Secondly, the spectral shift of electromagnetic resonances in spheroids due to *confocal* dielectric overlayers of finite thickness can be calculated with a formula derived in /365/. Corresponding results for prolate ellipsoids fit the experimental data quite well (solid line in Fig. 24a). The details of the calculation are presented elsewhere /267/. We only note here, that the dimensions of the ellipsoids (bumps) have to be $\approx 1 - 2$ nm in order to fit the experimental data. This is a consequence of the fast saturation of the resonance shift with pyridine overlayer thickness. Following /175/, we estimate a decrease of the electromagnetic enhancement by $\gtrsim 10$ for the second layer of adsorbed molecules compared to the first. Coldly evaporated silver films exhibit a *short-range electromagnetic enhancement* (as assumed in /366/) in contrast to some other silver surfaces investigated (e.g. /255/ and Sect. 4.3).

Thirdly, the coverage dependence of the intensity at resonance (Fig. 24b) presumably reflects the density of SERS active molecules on the silver surface /267/. What are SERS active molecules? As mentioned earlier, pyridine adsorbs in two configurations on silver /20/: a low coverage, essentially π-bonded species (phase I), and a high coverage, nitrogen lone-pair bonded species (phase II). As we can trace the SER signal of pyridine down to very small exposures /99/, π-bonded molecules on certain active adsorption sites as discussed in, for instance, /239/ must be identical with surface pyridine (see also Sect. 4.4; note, that only part of phase I molecules constitutes the SERS active species). These are subject to the *full* enhancement ($\approx 10^4$, Fig. 16), the electromagnetic ($\lesssim 10^2$; E_{bulk} in Fig. 16) as well as the chemical contribution ($\gtrsim 10^2$; E_{chem} in Fig. 16). If we assume that phase II molecules feel essentially only the electromagnetic enhancement and show ν_1 at 993 cm^{-1} like bulk pyridine, the data in Fig. 24b mirror the exposure dependence

of the density of surface pyridine molecules (see also /268/). This density may be
estimated in a simple approach under the following assumptions /267/: (i) the sticking probability for an incident molecule is unity; (ii) on clean parts of the surface molecules adsorb as phase I species; (iii) a molecule, which adsorbs on a surface pyridine covered part of the surface, either starts to build the second layer
[probability (1 - S)] or squeezes into the first layer and adsorbs as phase II species
[probability S; (1 - S) accounts for the fact, that even for multilayer coverage we
still must have some surface pyridine molecules (Fig. 24b); these might be molecules
on *selected* SERS active sites (details unknown)]; (iv) when a molecule squeezes
into the first layer, it moves a second, phase I molecule into the upright orientation (hence the density of phase II is twice that of phase I; see also /20/). The
differential equation, which describes the development of the fraction of saturation coverage of surface pyridine molecules (Θ) as a function of integrated exposure (E), is then given by /267/:

$$\Theta(E) = (1 - S)\{1 - [1 - ES/(1 - S)]exp(-E)\} \quad . \tag{6}$$

For S = 0.84 (solid) and S = 0.95 (dotted) numerical results are displayed in Fig.
24b. The curves fit the experimental data quite well for sub-monolayer coverages.
The agreement is poor for high coverage, probably due to a change of the resonance
properties neglected here. Several further observations can be taken as support /267/
for the picture described by (6), for instance the splitting of the N 1s peak for
pyridine exposed silver foils as measured by XPS /260,367/ and its relation to SER
features. Note, that *"ordinary"*, π-bonded pyridine, i.e. molecules on flat parts
of the surface, does not contribute to low coverage SER spectra in our interpretation which is in agreement with corresponding results from Ag(111) /268/.

Although the electromagnetic model explains most features of SER excitation profiles quite reasonably, there remain difficulties. Main problem is, that optical
properties of very small bumps ($\approx$ 2 nm dimensions) are only crudely described by
continuum electromagnetic procedures and that our interpretation allows an only moderate wavelength dependence of the chemical contribution to the enhancement. Moreover, the mode specific spectral dependence $I_{bulk}/I_{surface}$ (Fig. 25) still needs
to be explained. Alternatively, an interpretation of the measured excitation profiles based on optical excitations involving charge transfer between pyridine and
localized surface electronic states /368/ has been proposed /67,264/. In closing
this section we note, that recently published SER excitation profiles from coldly
evaporated films on an island structure /94/ and from films condensed at 15 K /369/
are in fair agreement with the results presented here.

4.1.5 Comparison of Results from Various Experiments

In Fig. 27 we compare the SER features in the region of the strong breathing vibrations from pyridine on various coldly evaporated silver films. Mono- as well as multilayer spectra from thin Ag films laid down on a silver optical grating (Fig. 27a) are similar to the corresponding spectra from coldly evaporated films displayed in Fig. 8. They are, however, considerably less intense than equivalent measurements at *thick* films (Figs. 8 and 27e, /50/; the spectra shown in Fig. 27a have been recorded by resonantly exciting plasmon surface polaritons at the periodically corrugated surface). Figure 27b shows SER spectra from thick Ag films deposited on quartz substrates at 13 K, where part of the substrate had been coated by a silver island structure. Weak signals from ≈ 3 layers of pyridine have been observed after annealing the exposed sample to ≈ 70 K /94,370/ [this so called "low temperature anneal" /370/ was necessary to see the full peak strength; the effect has been attributed

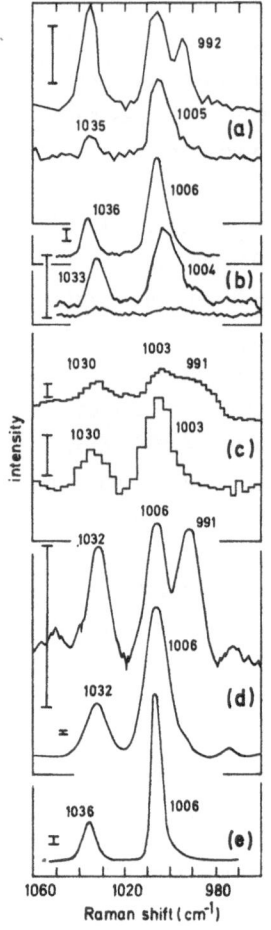

Fig. 27. Comparison of SER spectra from pyridine on various coldly evaporated films.

(a): 2 nm of Ag evaporated on silver grating (Λ = 800 nm, h = 50 nm) at 120 K; coverage of 0.5 (bottom) and 3.5 monolayers of pyridine; 50 mW of 514.5 nm radiation, 8 cm^{-1} bandpass; after /50/;

(b): 30 nm of Ag deposited at 150 K; lower curves: exposed at 13 K and "low temperature annealed" at 70 K (bottom: film on quartz substrate, top: film on Ag island structure); upper curve: film on Ag island structure after "high temperature anneal" at 200 K, recorded at 150 K; $18 \cdot 10^{14}$ molecules per cm^2 coverage; 150 mW of 530.9 nm radiation; after /94/;

(c): lower spectrum: 15 nm of Ag evaporated on silver substrate at 180 K and exposed to 2 L (uncorrected) of pyridine; 300 mW of 488.0 nm radiation, 7 cm^{-1} bandpass; after /271/; upper spectrum: 50 nm of Ag deposited on copper substrate at 100 K, 5 ± 1 monolayers of pyridine coverage; 150 mW of 488.0 nm radiation, 7 cm^{-1} bandpass; after /366/;

(d): thick Ag film deposited on polished copper substrate at 15 K, condensation of thick (250 - 1000 nm) pyridine overlayer; top: sample at 15 K, bottom: sample annealed to 200 K; 250 mW of 647.1 nm radiation, 5 cm^{-1} bandpass; after /369/;

(e): thick Ag film evaporated on polished Cu substrate at 120 K, 0.2 L of pyridine exposure; sample annealed to 210 K; 200 mW of 514.5 nm radiation, 4 cm^{-1} bandpass; after /358/.

The bars to the left of the spectra represent 10^2 cts/s, except for (e), where it represents 10^3 cts/s

to thermally stimulated, irreversible movement to and/or reorientation of pyridine molecules at active sites /370/; annealing of defect sites in the bulk of evaporated films might also contribute via electromagnetic enhancement mechanisms, which are stronger for less disturbed layers (see Chapt. 3)]. It is somewhat surprising, that, besides the surface pyridine line, only a weak bulk pyridine line at ≈ 990 cm^{-1} is observed (overestimation of coverage?). After a "high temperature anneal" /94/ the spectra are further enhanced (Fig. 27b) similar to those discussed in Sect. 4.1.3. Within experimental accuracy, peak positions of ν_1 and ν_6 are identical to those of pyridine on thick coldly evaporated Ag films laid down at 120 K and annealed to 210 K (Fig. 27e). Figure 27c shows spectra from roughly a monolayer of pyridine on a thin Ag film deposited at 180 K /271/ and from 5 ± 1 layers on a thick film deposited at 100 K /366/. Although spectral features are only weakly pronounced, the additional line at ≈ 991 cm^{-1} for the thicker coating is clearly seen. Finally, thick pyridine layers (250 - 1000 nm) laid down on thick silver films deposited on a polished copper block at 15 K /369/ exhibit spectra similar to samples prepared at 120 K (compare Fig. 27d to Fig. 27e and Fig. 8). Annealing to 200 K (desorption of multilayer pyridine) has a similar effect as discussed in Sect. 4.1.3. SERS from thick pyridine layers on silver films deposited and exposed at liquid He temperature has been reported earlier /133/. Here spectral features of surface pyridine are very similar to corresponding results from Ag deposited at 120 K (see Table 2 and Table 3 in /133/).

In summary, thick coldly evaporated silver films seem to be stronger enhancers than thin films. For sub-monolayer pyridine coverage, SER spectra are dominated by a species with ν_1 at 1003 - 1006 cm^{-1}, and ν_6 at 1030 - 1036 cm^{-1} is relatively weakly pronounced. A third line at 991 - 993 cm^{-1} develops with increasing coverage. The intensity of ν_6 grows faster with coverage than that of ν_1 and matches or exceeds the latter for a coverage of several layers. These fairly consistent experimental data can be understood within the frame outlined in the preceeding sections, i.e. can be interpreted in terms of strongly enhanced surface and weakly enhanced bulk pyridine contributions.

For the sake of completeness we briefly mention investigations on films evaporated or sputtered and exposed at room temperature /153,157/. Spectra of rather low quality exhibit weak features at ≈ 1009 cm^{-1} and ≈ 1035 cm^{-1} (similar features are observed from Pt, Pd, Ti, or Ni films /154/, where Ni shows a much stronger signal than Ag!). To the opinion of the author, careful cross checks of the experimental conditions are necessary before any sound conclusion can be drawn from these somewhat unique results.

4.2 Coldly Evaporated Copper and Gold Films

SER spectra from various coldly evaporated noble metal films (group Ib) exposed to
0.2 L (Ag) and 2 L (Au,Cu) of pyridine are displayed in Fig. 28 /123/. Corresponding
line intensities from silver and copper samples are comparable, whereas the gold
sample exhibits only ν_1 with a roughly 30 times smaller intensity. In fact, after
evaporation and exposure we did not observe any line from Au. The spectrum shown in
Fig. 28 has been recorded after warming the sample to 210 K and recooling to 120 K,
a procedure known to increase SER intensities from Ag /100/ (very recently, a somewhat more intense spectrum from Au films has been reported /369/; like Cu, but unlike Ag, the "quality" of SERS active Au films depends critically on evaporation
conditions; see /273/ for Cu). The relative SER line intensities of pyridine on Cu

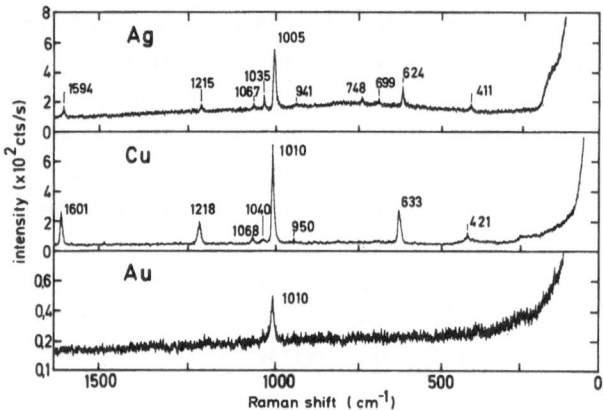

Fig. 28. SERS from pyridine
on coldly evaporated Ag
(0.2 L), Cu (2 L), and Au
films (2 L). 60 mW of
676.4 nm radiation, 4 cm^{-1}
bandpass. Au film has been
warmed to 210 K and recooled
to 120 K before measurement.
After /123/

are different from those on Ag as is easily seen from the intensity ratio of ν_4
($\approx$ 1600 cm^{-1}) and ν_1 ($\approx$ 1000 cm^{-1}, Fig. 28).

SER line positions and relative intensities of pyridine on Cu and Ag films are
compared to corresponding data from electrode surfaces in Table 3 /123/. Spectral
features are almost identical for Cu films and electrodes. Compared to silver,
lines from pyridine on Cu films are only slightly shifted to greater energies, which
suggests similar bonding on both metals. Interestingly, SER data from Cu samples
approach those of the copper pyridine complex. For red excitation, line intensities
of high energy modes (e.g. ν_4) are more pronounced for copper than for silver. However, exciting silver samples with green or blue radiation (Table 3 and Fig. 9),
relative line intensities similar to those from Cu are observed (red light excitation). Changes of relative line intensities with excitation wavelength have been
explained with mode specific excitation profiles (Sect. 4.1.4 and /119,360/). The
shift of the resonance-like profiles to greater excitation energy with vibrational

Table 3. Vibrations of pyridine in various systems. (a): on activated silver electrode at -0.6 V_{SCE}, 647.1 nm excitation, after /123,371/; (b): on SERS active silver film in UHV (0.2 L exposure), 676.4 nm excitation, after /123/; (c): complexed pyridine, $Cu(Py)_2Ni(CN)_4$, IR study, after /347/; (d): on activated copper electrode at -0.6 V_{SCE}, 647.1 nm excitation, after /123,371/; (e): on SERS active copper film in UHV (2 L exposure), 676.4 nm excitation, after /123/

Mode Symmetry	(a) SERS from Ag Electrode C_5H_5N	(b) SERS from Ag Film C_5H_5N	(c) IR Complex $Cu(Py)_2Ni(CN)_4$	(d) SERS from Cu Electrode C_5H_5N	(e) SERS from Cu Film C_5H_5N
ν_{21}, A_2	391 (5)			385 (3)	
ν_{27}, B_2	419 (12)	411 (7)	435 (m)	422 (14)	421 (9)
ν_3, A_1	635 (55)	624 (27)	640 (m)	635 (67)	633 (32)
ν_{12}, B_1	651 (9)		650 (w)	652 (12)	
ν_{26}, B_2	698 (6)	699 (7)	689 (vs)	699 (9)	701 (2)
ν_{23}, B_2	753 (4)	748 (6)	753 (s)	755 (4)	
ν_{25}, B_2	872 (3)		868 (vw)		
ν_{20}, A_2					
ν_{24}, B_2	943 (6)	941 (5)	949 (m)	946 (6)	950 (2)
ν_{22}, A_2					
ν_1, A_1	1013 (100)	1005 (100)	1017 (s)	1013 (100)	1010 (100)
ν_6, A_1	1036 (20)	1035 (19)	1043 (s)	1041 (19)	1040 (4)
ν_8, A_1	1067 (17)	1067 (7)	1068 (s)	1067 (31)	1068 (8)
ν_{17}, B_1			1088 (w)		
ν_{16}, B_1	1157 (3)		1154 (s)	1156 (7)	
ν_5, A_1	1216 (34)	1215 (8)	1219 (s)	1218 (64)	1218 (21)
ν_{11}, B_1			1241 (s)		
ν_{15}, B_1			1360 (w)	1359 (2)	
ν_{18}, B_1	1449 (2)		1449 (vs)	1448 (8)	
ν_9, A_1	1486 (3)		1487 (s)	1486 (8)	1485 (2)
ν_{14}, B_1	1574 (3)		1575 (m)	1571 (8)	1573 (2)
ν_4, A_1	1602 (46)	1594 (11)	1609 (vs)	1603 (77)	1601 (33)

energy causes high energy lines to be less prominent for red excitation (Ag). As the excitation profile maximum for Cu is shifted to longer wavelength with respect to Ag (Fig. 22, /123/), one might expect qualitatively similar relative intensities for Cu (*red* excitation) and Ag samples (*green, blue* excitation), as one works in either case on the high energy side of the excitation profile resonance. Indications for this trend can be extracted from Table 3. Note, that weak signals from pyridine on Au and Cu have also been detected with green excitation, but not with blue /123/. Another remarkable feature is the rather low intensity of ν_6 from Cu samples (Fig. 28 and /123/) which is not understood at present. A recent comparative SER study of pyridine on Ag, Cu, and Au deposited at 15 K /369/ has confirmed the results from /123/ outlined above.

Recently, coverage dependence and annealing behaviour of SER features from pyridine on coldly evaporated Cu films have been studied in detail /273/. Qualitatively, similar effects as with silver have been found (Sects. 4.1.2 ,3). However, spectral features are more stable against annealing. This might be explained with the lower mobility of Cu surface atoms compared to Ag /274,275/. In fact, Cu films laid down at room temperature or samples annealed to/at room temperature /273,369/ may still be SERS active. Figure 29 displays Raman spectral features from pyridine on films deposited at ≈ 130 K, at ≈ 230 K, and at room temperature, respectively. All films have been exposed and measured at 120 K. In contrast to coldly evaporated films, samples prepared at room temperature display pronounced bulk pyridine peaks (within experimental accuracy at the same energy as for silver). The spectra displayed in Fig. 29 mirror presumably the different surface topography and bulk pro-

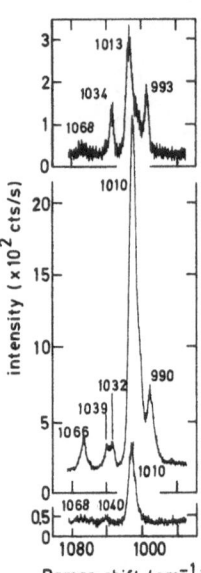

Fig. 29. SER spectra from Cu films deposited at ≈ 130 K (bottom), ≈ 230 K (central), and at room temperature (top). All samples are exposed to 2 L of pyridine at 120 K and measured at this temperature. 190 mW of 647.1 nm radiation, 4 cm^{-1} bandpass. As spectra are from different experimental runs, absolute intensities cannot be compared. After /273/

perties of films prepared at different temperatures. The magnitude of both, electromagnetic and chemical contributions to SERS, change with deposition temperature. The
latter, because point defect related, SERS active sites (Sects. 4.1.4 and 4.4) anneal at higher temperatures, the former, because the density of bulk defects is
smaller and the shape and density of surface bumps might be more favourable for
films prepared at higher temperature (see also Sects. 4.1.3 , 4). Obviously, Cu samples prepared at $\approx$ 230 K give the best SERS performance: both, surface (chemical
and electromagnetic enhancement) and bulk pyridine lines (only electromagnetic enhancement), are very strong. A somewhat more detailed discussion may be found elsewhere /273/.

Like Cu, gold samples may exhibit enhanced Raman signals at room temperature
/369/. Here more experimental work is necessary. Besides the noble metals of group
Ib, only coldly evaporated sodium /142/ and lithium /133/ films (see Chapt. 5) as
well as aluminum samples /144/ display SERS. The latter does not show, however, any
characteristic line after pyridine exposure /144/.

4.3 Surfaces Prepared with Various Techniques

4.3.1 Silver

Raman spectral features in the breathing mode region for pyridine on various silver
surfaces are compared in Fig. 30. Approximately three layers of pyridine deposited
on island films at 13 K and annealed to 70 K ("low temperature anneal", /94/) display bulk (990 cm^{-1}) and surface (1000 cm^{-1}) pyridine ν_1 peaks with approximately
the same intensity. The antisymmetric stretching vibration ν_6 is almost as strong
as ν_1, which indicates bulk as well as surface contributions to this line. Only
about 150 cts/s peak intensity have typically been observed. Upon annealing to 200 K,
all features except a small peak at $\approx$ 1006 cm^{-1} disappear /94/. For thin as well as
thick layers on silver optical gratings no surface pyridine line is seen, and ν_6 is
rather weakly pronounced (Fig. 30b, /48/). The spectra have been taken by resonantly
exciting plasmon surface polaritons. The coverage dependence of the Raman features
points to a pronounced first layer effect. Signals from the first layer are $\approx$ 100
times stronger than those from subsequent layers /48,268/. Note the overall weakness
of the signals (peak intensity $\approx$ 50 cts/s). The enhancement of Raman bands from
pyridine on iodine roughened Ag (Fig. 30c) has been attributed to long range electromagnetic effects caused by surface roughness features of $\approx$ 50 nm lateral dimension
/98/. Two different adsorbed states corresponding to the first layer and succeeding
layers, respectively, with overlapping ν_6 are believed to contribute to the spectra.
Again, note the weakness of the spectra with peak count rates of $\approx$ 50 cts/s. The
sputter-cleaned silver surfaces used in /97/ also lead to relatively weak signals.

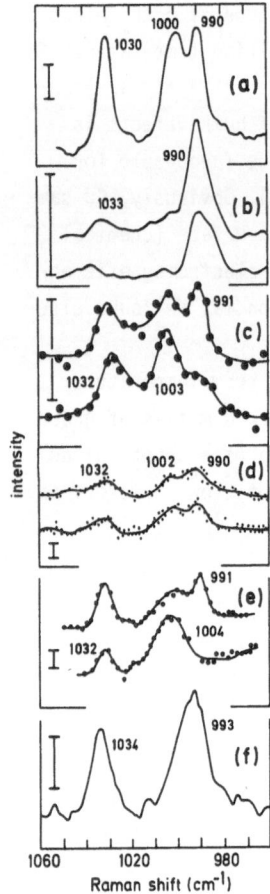

intensity

1060   1020   980
Raman shift (cm⁻¹)

Fig. 30. Comparison of Raman spectra from pyridine on various silver substrates.

(a): ≈ 3 layers on island film (≈ 100 nm lateral dimensions); 150 mW of 530.9 nm radiation; after /94/;

(b): ≈ 1 (bottom) and ≈ 25 (top) layers on periodically corrugated Ag(111) surface (Λ = 1000 nm, corrugation depth ≈ 100 nm); 150 mW of 514.5 nm radiation, 6 cm⁻¹ bandpass; after /48/;

(c): ≈ 1 (bottom) and ≈ 4 (top) layers on photochemically roughened silver surface; several hundred mW (?) of 488.0 nm radiation, 8 cm⁻¹ bandpass; after /98/;

(d): ≈ 0.3 (bottom) and ≈ 1.2 (top) layers on sputter-cleaned Ag foil; 100 mW of 514.5 nm radiation, 6 cm⁻¹ bandpass; after /97/;

(e): ≈ 1.5 (bottom) and ≈ 10 (top) layers on "smooth" Ag(100); 100 mW of 514.5 nm radiation, 5.5 cm⁻¹ bandpass; after /101/;

(f): ≈ 1 monolayer on smooth Ag(111); 1 W of 514.5 nm radiation, 10 cm⁻¹ bandpass; after /372/.

The bars to the left of the spectra represent 50 cts/s (a - d) and 1 ct/s (e,f)

Sub-monolayer and ≈monolayer spectra do not differ very much. Both exhibit ν_1 peaks at 990 cm⁻¹ as well as at 1002 cm⁻¹ along with a ν_6 line at 1032 cm⁻¹ of somewhat smaller intensity. All features disappear upon annealing the sample to ≈ 200 K /97, 269/. The two ν_1 lines have been interpreted as being due to pyridine adsorbed to two different sites. Multilayer Raman scattering was not observed below an exposure of 330 L (uncorrected). Finally, mono- and multilayer spectra from pyridine on Ag(100) (Fig. 30e, /101/) are qualitatively similar to results from chemically roughened samples (Fig. 30c). The intensity is, however, much smaller (≈ 2 cts/s peak intensity!). The enhancement for the surface pyridine line has been estimated to ≈ 400 /101/. These results disagree with recent experiments on Ag(111), Ag(110), and Ag(100) surfaces (Fig. 30f, /372/). Here only a *single* ν_1 peak at 993 cm⁻¹ along with ν_6 at 1034 cm⁻¹ has been observed. The Raman intensity increases linearly from sub-monolayer to multilayer coverage and the depolarization ratio is low. This points to *ordinary* Raman scattering /372/. It has been argued /28/, that the line at 1004 cm⁻¹ in the spectra of /101/ (Fig. 30e) is caused by pyridine on special sites, which are usually not available on carefully prepared single crystal surfaces (presumably defect sites like steps, kinks, adatoms or vacancies as suspected in /372/). A comparison of the ordinary spectrum from Ag(111) (Fig. 30f) to the spectra displayed in Fig. 30a - d suggests a weak enhancement of $\lesssim 10^2$ for the latter (note, that this crude estimation neglects differences in the experimental procedures).

The SER spectra (a) - (d) in Fig. 30 have some features in common. They display, even for monolayer or less coverage, either both, the bulk as well as the surface pyridine line, or only the bulk pyridine ν_1 signal. The bands are roughly two orders of magnitude weaker than corresponding lines from thick coldly evaporated films (Fig. 27e), and the strong domination of the surface pyridine line for low coverage is not observed. A small density or absence of SERS active sites and hence of surface pyridine in combination with electromagnetic enhancement due to surface roughness (grating, island) can qualitatively account for the observed features. The comparison of Fig. 27 and Fig. 30 provides additional evidence, that the strong enhancement (electromagnetic plus chemical) is restricted to surface pyridine only.

4.3.2 Other Materials

The report of appreciably enhanced Raman signals ($\approx 10^5$) from pyridine on a drop of mercury /149/ has recently caused much excitement. Because of its optical properties and its smooth surface, no electromagnetic enhancement is expected from mercury. Spectra from the drop in gaseous or liquid pyridine or benzene have been compared to spectra recorded without the mercury drop. With mercury present, about 20 times larger intensity of ν_1 has been observed, from which the enhancement given above was estimated (ν_1 is found at 992 cm^{-1}; for liquid benzene the intensity is only doubled with the drop present). There are, however, several unsuccessful attempts by other groups to reproduce the results /67/. It is the opinion of the author, that the results reported in /149/ are either experimental artefacts and/or are misinterpreted. They should not be included into the general SERS discussion.

In /153,154,157/ Pd, Pt, Ti, or Ni films evaporated or sputtered at room temperature have been exposed to saturated pyridine vapour for 1 h, evacuated for 30 min at 10^{-5} Torr, and investigated. Weak features at pyridine frequencies in Raman spectra of rather moderate quality have been interpreted as surface enhanced lines. We think, that the presented data do not justify this conclusion. More experiments to separate ordinary Raman contributions along with a detailed, quantitative evaluation of intensities is necessary for sound conclusions. Like the mercury result, the observations reported in /153,154,157/ should not enter the general discussion of SERS at present.

4.4 Discussion and Conclusions

From the rich body of SER studies of pyridine adsorbed to metal/vacuum interfaces several conclusions can safely be drawn. *Three* species can be distinguished in the spectra from silver, which is best illustrated with the symmetric breathing vibration ν_1. (i) Surface pyridine displays ν_1 at 1003 - 1006 cm^{-1}. Only this species is subject to the strong enhancement ($\approx 10^4$ for coldly evaporated silver films), which is composed of an electromagnetic ($\lesssim 10^2$) and a chemical ($\gtrsim 10^2$) part. Surface pyridine are molecules chemisorbed to certain SERS active sites at the silver surface (see below). Bulk pyridine consists of two species. (ii) Molecules bonded to SERS inactive parts of the metal surface display ν_1 at 990 - 993 cm^{-1}, which indicates physisorption. These feel essentially only the electromagnetic part of the enhancement (a *weak* chemical contribution cannot be excluded). (iii) ν_1 is found between 991 - 996 cm^{-1} for multilayer pyridine, which overlaps the range given under (ii). There is some evidence, that ν_1 of multilayer pyridine is slightly shifted to greater energies compared to (ii) /358/. Multilayer pyridine exhibits either only ordinary Raman scattering or weak SERS caused by *long* range electromagnetic effects (gratings, island films). Short range electromagnetic enhancement as observed from coldly evaporated films is essentially restricted to (ii) and, of course, (i). We take the opportunity to emphasize, that, following common usage, the label "SERS active" for a site or adsorbed molecule refers to the chemical contribution to SERS. SERS inactive molecules (molecules on inactive sites) may well display Raman features surface enhanced by electromagnetic effects as outlined.

It follows, that *intense* SER spectra can only be expected from substrates, which simultaneously exhibit (i) a high density of SERS active sites, (ii) appropriate optical properties (i.e. small damping), and (iii) suitable surface topography to support electromagnetic resonances. This is reflected by the fact, that SERS so far *has convincingly been demonstrated only for appropriately prepared metals of high reflectivity* (Ag, Au, Cu, Li, Na). Coldly evaporated films obviously give the best performance: these meet apparently (i) to (iii) in a unique way. On the other hand, silver gratings or island films give rise to weaker SER spectra featuring mainly bulk pyridine: the density of SERS active sites is small on these surfaces, whereas (ii) and (iii) are matched.

On silver, SERS active sites are stable only at low temperatures ($\lesssim 220$ K). They anneal at room temperature. Moreover, SERS activity depends on the adsorbed molecule, i.e. is *molecule specific*: methane and ethane on otherwise SERS active surfaces do not show SERS /133/. Certain bonding properties or adsorption geometries to the active sites are presumably necessary for the chemical effect in SERS. In other words, SERS activity is a property of the entire complex, adsorbed molecule plus active site. Therefore, SER features may change with, for instance, pyridine coverage, because bonding properties and hence the density of SERS active surface pyridine may change, although the density of active metal sites is not affected (Sect. 4.1.2).

For coldly evaporated films, the electromagnetic contribution to SERS is responsible for the observed excitation profile resonances. As expected within this interpretation, the resonances shift with coverage to the red and upon annealing slightly to the blue. Note, that there is room for only a rather flat wavelength dependence of the chemical contribution to SERS within this concept.

Variations of SER spectral features and intensities for differently prepared surfaces (Figs. 27, 30) are caused by several effects. The electromagnetic and chemical share of the total enhancement depends on surface preparation. The density of SERS active sites is influenced by the surface treatment, and the scale of supra-atomic roughness features and hence the range of the electromagnetic enhancement is different for different surfaces. The remarkable decrease of SER intensities from surface pyridine on coldly evaporated films for exposures $\gtrsim 0.5$ L (Fig. 16) is due to two effects, the red shift of electromagnetic resonances with coverage, which is also responsible for the slight decrease of the bulk pyridine signal from physisorbed molecules, and the change of the surface pyridine density (Sects. 4.1.2 , 4; the density of active metal surface sites does not change). The annealing behaviour is more complicated: desorption of molecules may be involved. The increase of the surface pyridine signal between 180 K and 220 K is presumably mainly caused by physisorbed pyridine molecules, which start to migrate at the surface and either finally desorb or are trapped at vacant active sites in this temperature range. The latter process increases the surface pyridine density. Other contributions to the increase of the surface pyridine signal may come from the blue shift of the excitation profile resonance with temperature (Fig. 26) and an orientation of surface pyridine to more favourable adsorption geometries (hindered at low temperature and/or by neighbouring molecules). Several effects contribute to the decrease of the SER signal for $T \gtrsim$ 210 K (Fig. 26). Most likely, annealing of active sites plays a leading role. Annealing of supra-atomic scale roughness (bumps), slow desorption of chemisorbed pyridine, or change of bonding properties may also contribute. In closing this part we note, that similar conclusions have recently been drawn in another paper /373/.

It remains an important open question: what is the nature of SERS active sites? There is strong evidence that atomic scale roughness is involved /67/. As coldly evaporated surfaces are poorly defined, it is, however, difficult to extract details from experiments. Early attempts to demonstrate the importance of atomic scale roughness by deposition of $\lesssim$ monolayer amounts of silver at low temperature on inactive silver surfaces were unsuccessful /271,373,374/. Studies of appropriately prepared single crystal faces with known defects and defect densities seem to be more promising. Such experiments are just about to start /372/. Despite appreciable lack of knowledge, available SER vibrational data and UPS results /20/ may allow some preliminary conclusions on the pyridine/silver system. The ideas contain, however, some degree of speculation and have to be cross checked by suitable experimental studies.

Coldly evaporated silver films have presumably a (111) fiber texture /277,375/ with various kinds of surface defects. Figure 31 schematically displays a surface with different imperfections. These include irregularities like adatoms, vacancies, clusters, kinks, steps, and dislocations and adsorbed or incorporated impurities.

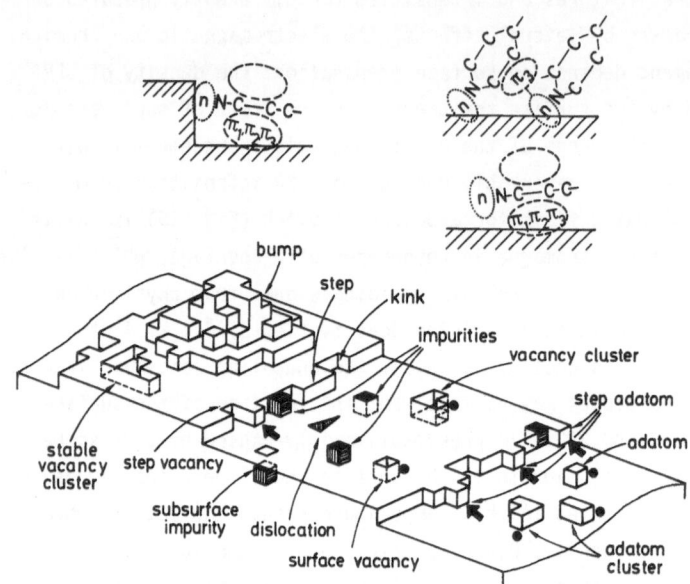

Fig. 31. Sketch of irregular surface with various types of defects. For silver, defects marked by a dot are annealed at temperatures for which the surface is still SERS active. Defects, which presumably are important for SERS from pyridine on Ag, are marked by an arrow. Upper diagrams schematically show proposed orientation and bonding of pyridine to Ag(111) [low coverage (bottom) and high coverage compressional phase (top), right hand diagrams; after /20/] and to defect sites on coldly evaporated silver (low coverage, left hand diagram; because of the local charge distribution due to the Smoluchowski-effect /377/, an inclination of the molecule as indicated seems reasonable).

In addition to point defects, roughness on a supra-atomic scale might be present (bumps as indicated in the left part of Fig. 31). Some surface defects are instable at those temperatures, for which the silver surface is still SERS active. Adatoms, vacancies, and *small* clusters (marked by a dot in Fig. 31) anneal at temperatures below 210 K /67,100,288,289,376/. They are therefore not important for SERS from coldly evaporated Ag films. Depending on the identity of the adsorbed species, this may or may not be true also for impurity defects. Due to the smoothing effect of the metal electron charge distribution in the topmost surface layer, surface defects on metals are usually positively charged /377/ (more exactly, this smoothing effect leads to localized dipole moments and large local electric fields /378,379/, which are responsible for, e.g., the decrease of the work function on surface with increasing defect density /380,381/ or increased C-H bond-breaking abilities of low coordination number sites on Pt /320/). It is known, that the unique local structural environment and charge density of defect sites may lead to enhanced chemisorption kinetics /320,382 - 384/ and changes in binding states and structure of adsorbates /384/. Similar effects may be expected in the vicinity of adsorbed electronegative impurities like oxygen which induce a positive charge on neighbouring surface metal atoms. Adsorption of certain molecules may take place selectively on these positively charged sites. The strong promotion of ethylene adsorption on Ag(110) by dissociatively adsorbed oxygen has been interpreted in this way /385/.

Without knowing details, we tentatively assume that pyridine preferentially bonds to defect sites with appreciable dipole moment. In addition to the general arguments given in the preceeding section these sites are favoured because of the large dipole moment of pyridine: dipole-dipole attraction contributes to the bonding strength. Based on the known adsorption geometry of pyridine on smooth Ag(111) /20/, we suggest an orientation and bonding for pyridine chemisorbed to defect sites as sketched in Fig. 31 (on terrace sites we assume physisorption as described in /20/). With respect to SERS from Ag, only sites still stable at 210 K, but instable at room temperature, are important. These are essentially defect sites at steps (e.g. adatoms at steps /100/; fat arrows in Fig. 31). Consequently, we identify the species bonded to these sites with surface pyridine /100,267/ (ν_1 at 1006 cm^{-1}; bonding to as yet unspecified impurity sites is presumably responsible for the line at 1026 cm^{-1} as outlined in Sect. 4.1.1). As the bonding geometries of surface pyridine and suitably complexed pyridine /347,359/ display some similarities, we expect similar vibrational properties, which is indeed observed (Tables 2 and 3; the somewhat smaller blue shift of surface pyridine lines (ν_1, ν_3, ν_4) compared to complexed pyridine might partly be caused by the additional bonding through the π orbitals, see Fig. 31). Due to the different local environment of defect sites at steps (Fig. 31), surface pyridine lines are expected to be inhomogeneously broadened. Under certain conditions this may lead to the asymmetric line shapes observed in SER spectra (Fig. 12; a more detailed discussion of this point is given in Chapt. 5). For increasing coverage, adjacent adsorbed molecules disturb the adsorption geometry and bonding properties of surface pyridine. Therefore the density of surface pyridine may be reduced /267/ and/or the efficiency of the chemical effect attenuated /386/, when the SERS active species is embedded into the high coverage compressional phase /20/. Within this frame, spectral changes of strongly enhanced surface pyridine signals with coverage, upon annealing, and with surface preparation can qualitatively be understood, if the influence of supra-atomic roughness as discussed in Sect. 4.1.4 is considered as well. In passing we note, that the density of defect sites is particularly high at bumps of small size (Fig. 31).

The nature of the chemical enhancement is still unclear (Sects. 2.2 , 3). It is unclear as well, why only pyridine bonded to certain defect sites displays strong chemical enhancement. More information on the local electronic structure and charge density, i.e. on the unique features of the bonding to SERS relevant defect sites, is urgently needed. UPS studies of pyridine adsorbed to appropriately prepared surfaces with defined defect structure could help to clear the situation. With respect to photon driven charge transfer models for the chemical enhancement /67,236/, it is conceivable, that bonding to defect sites leads to the electronic structure optimizing this process. Note, that this may be specific for pyridine adsorption and not necessarily a general rule for the chemical contribution to SERS. As to chemical effects in general, one should also recall, that charge transfer from the metal is facilitated at positively charged sites, i.e. at defect sites /382/ (independent of the adsorbate). At the present state of knowledge, any final conclusion would be, however, premature. A detailed knowledge of the local electronic structure of the pyridine/metal complex is necessary to unravel the chemical effect. In closing we again point out, that low coverage pyridine SER features are only from molecules adsorbed to selected, SERS active sites of atomic scale roughness. Interestingly, defect sites play a central role in heterogeneous catalysis /320,321/. SERS may therefore become a helpful method for the study of catalytic processes on certain surfaces /67/ (see also Chapt. 10).

5. Hydrocarbon Adsorption

Hydrocarbons play a central role in various heterogeneously catalyzed reactions of industrial importance /387/. Metallic catalysts are frequently used in these reactions /388,389/. Examples are the hydrogenation of benzene to cyclohexane over supported nickel, palladium, or platinum /387/, the selective hydrogenation of acetylene to ethane (nickel or palladium catalysts /387/), the hydrogenolysis of C-C bonds by nickel, and the complete oxidation of hydrocarbons by palladium and platinum /389/. Although less widely used than metals of group VIII, the noble metals of group Ib display some unique features in heterogeneous catalysis. The selective oxidation of ethylene to ethylene oxide is catalyzed by supported silver /390 - 393/, and alcohols are oxidized to aldehydes or ketones by silver or copper catalysts /389/. Studies of hydrocarbon adsorption on metal surfaces are of fundamental interest for an approach to the understanding of many catalytic processes. In this chapter we summarize SER vibrational studies of hydrocarbon adsorption on mainly silver. The data are compared to results obtained with other techniques.

5.1 Open-Chain Hydrocarbons on Silver

5.1.1 Alkanes

Saturated hydrocarbons like methane and ethane interact only weakly with metal surfaces at low temperature. The small heat of adsorption of $\approx$ 5 kcal/mole points to only physical adsorption even on the quite reactive surfaces of rhodium (CH_4, /394/), tungsten (CH_4, /395/), nickel (CH_4, /396/), and ruthenium (C_2H_6, /397/). Desorption of the physisorbed species is observed around 100 K /395,397/. Consistent with these observations, infrared vibrational studies of ethane (deuterated) on Cu(110) /398/ and on Pt(111) /399/ at liquid nitrogen temperature yielded C-D stretching frequencies which were only slightly altered with respect to those of the free molecule.

Therefore we do not expect CH_4 or C_2H_6 to stick on coldly evaporated, SERS active silver films at 120 K. Indeed, no characteristic Raman lines have been obtained from exposed samples /400/. At liquid helium temperature, either species should, however, condense on the surface. Nevertheless, coldly evaporated Ag films exposed to CH_4 and C_2H_6 at 11 K /401/ or colloidal silver in a solid ethane matrix /402/ did not display any SER feature. These very interesting results clearly demonstrate the importance of *chemical* bonding of the adsorbate for SERS, i.e. the importance of a chemical contribution to SERS. Electromagnetic enhancement alone apparently does not lead to observable Raman features from the surfaces studied in /401,402/.

5.1.2 Ethylene

The planar symmetrical ethylene molecule has D_{2h} symmetry. Its molecular structure is well known /403,404/. The isolated molecule has twelve normal modes of vibration: six are Raman active ($3A_g + 2B_{1g} + 1B_{2g}$), five are infrared active ($1B_{1u} + 2B_{2u} + 2B_{3u}$) and one is silent ($1A_u$). Because of its centrosymmetric structure, ethylene is subject to the principle of mutual exclusion. Vibrational spectra of gaseous /405 - 408/, liquid /409/, solid /410 - 413/, and matrix-isolated ethylene /141/, as well as of coordinated ethylene /415 - 418/ and some matrix-isolated ethylenic complexes /419 - 422/ have been published (in this review we use the notation of /408/ for the vibrational modes). The structure of ethylene complexes of copper and silver has been characterized by electron spin resonance spectroscopy /423/. Numerous investigations have been dedicated to the electronic properties of ethylene /424 - 426/. UV-absorption starts at 7.1 eV /427/, electron impact spectra /428/ show an additional singlet-triplet transition at 4.4 eV /429/. Interestingly, matrix-isolated ethylene complexes of noble metals (group Ib) are coloured /422/. The colour has been attributed to a charge transfer band associated with the excitation of the unpaired electron and centered in the visible region of the optical spectrum [for instance at $\approx$ 2.15 eV for $Ag(C_2H_4)$].

There exists an enormous body of literature on ethylene adsorption on metals. In general, ethylene interacts strongly with transition metals of group VIII, and the molecule is appreciably perturbed upon adsorption. It rehybridizes close to sp^3 configuration and essentially forms a di-σ bond to, e.g. Pt(111) /430 - 433/, Rh(111) /434/, Fe(110) /435/, and Ni(111) /436/ at low temperature (there is some uncertainty concerning ethylene adsorption on Ni(111) /433/: early UPS studies suggested π bonding /437/). Weaker interaction and less perturbation of ethylene has been reported for adsorption on Pd(111) at low temperature /438/. When warming exposed metals of group VIII to room temperature, new species are usually formed: ethylidyne on Pd(111) /439,440/, on Rh(111) /441/, and on Pt(111) /442/ [conversion to ethylidene /430/ and a vinyl-like species /443/ has also been discussed for Pt(111)] or acetylenic species on Ni(111) /437,444 - 446/ (in /446/ evaporated films rather than single crystals have been used).

Far less studies have been dedicated to ethylene adsorption on metals other than those of group VIII. Besides rhenium /447,448/, only copper and silver recently attracted attention. On both, Ag(110) /385,449/ and Cu(100) /450/, ethylene adsorbs at low temperature without significant rehybridization with its plane parallel to the surface (EELS studies; IR vibrational results are presented in /451/). The molecule desorbs below room temperature without significant decomposition products remaining on the surface. The picture of weak molecule-metal interaction of π character is supported by photoemission studies /452 - 454/. With respect to corresponding SER investigations, two EELS results are of particular interest. Firstly, adsorption on Ag(110) is strongly promoted by the presence of atomic oxygen on the surface /385/. It has been concluded, that ethylene bonds selectively only to those silver atoms on which a positive charge has been induced by the oxygen atoms (ethylene-oxygen interaction on silver surfaces is investigated in, e.g.,/455 - 457/). Secondly, electron energy loss spectra (electronic) from ethylene on coldly evaporated silver films display a band centered at $\approx$ 2.8 eV /263/ in addition to those observed from the isolated molecule (see above). This band has been assigned to a charge transfer state of molecules bonded to sites of atomic scale roughness /263/ (similar to a corresponding pyridine result; see Chapt. 4 and /263/).

SER vibrational studies of ethylene on silver /133,135,400,402,458 - 461/ and copper /462/ have been performed by several groups. Adsorption on coldly evaporated films has been investigated most comprehensively, especially by the author and co-workers /459,460,462/. These results will be outlined in some detail and briefly compared to spectra of other groups /135,402,458/.

General SER Features. Figure 32 displays SER spectra from coldly evaporated silver films exposed to various ethylene isotopes. As multilayers are not formed under our experimental conditions, the dose of 36 L results in at most monolayer coverage. Most lines of the highly structured spectra can be assigned to ethylene skeletal vibrations (see below; they are marked with a dot in Fig. 32). The strongest bands are the symmetric scissors mode ν_3 at 1331 cm^{-1} and the C-C stretching mode ν_2 at 1595 cm^{-1} (C_2H_4; peak intensities: several thousand cts/s). As can be estimated by analogy to pyridine SER results, the signals of ν_3 and ν_2 are enhanced by at least the same factor as ν_1 of pyridine (see Chapt. 4). The analysis of the spectra reveals several further interesting features.

i) As in pyridine SER spectra, C-H stretching modes are only weakly pronounced (Fig. 33a). Corresponding lines of the deuterated species are somewhat more intense.

ii) The weak line at $\approx$ 1080 cm^{-1} is assigned to the usually silent CH_2 twisting mode ν_4 /133/ (Fig. 33b; ν_4 of C_2D_4 and $^{13}C_2H_4$ is unfortunately obscured by other bands). Almost all modes, which are only infrared active for the isolated molecule, show up in SER spectra.

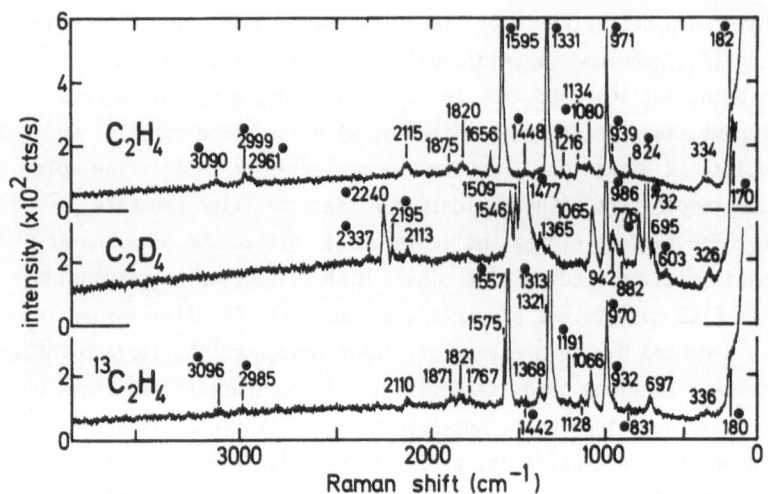

Fig. 32. SER spectra from coldly evaporated silver films exposed to 36 L of ethylene. Dots mark ethylene skeletal vibrations (see text). 200 mW of 514.5 nm radiation, 4.5 cm^{-1} bandpass. After /460/

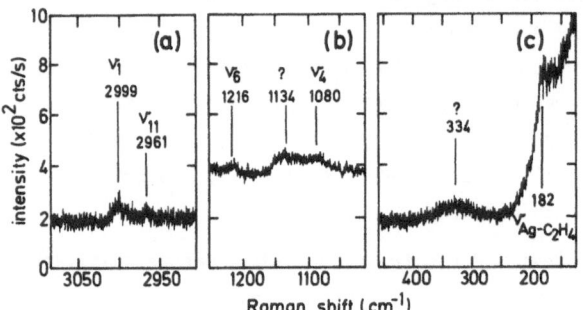

Fig. 33. Details of SER spectrum from Ag film exposed to 3.6 L (a and c) and 36 L (b) of C_2H_4. 200 mW of 514.5 nm radiation, 4.5 cm^{-1} bandpass. After /460/

iii) The weak feature at the slope of the Rayleigh line at ≈ 182 cm^{-1} (Fig. 33c) might be due to the metal-molecule stretching vibration.

iv) The strongest lines in SER spectra of C_2H_4 and $^{13}C_2H_4$, respectively, are assigned to ν_3 [peak intensity: $100(C_2H_4)/100(^{13}C_2H_4)$], $\nu_2(69/58)$, and to the symmetric wagging vibration $\nu_7(25/18)$. All other lines are considerably weaker (≲ 2). For the deuterated species, ν_2 is the strongest line (100), and ν_3 and ν_7 display similar intensity (25). It has been pointed out /133/, that the strongest bands in SER spectra of ethylene as well as of other alkenes and alkynes (Sect. 5.1.4) involve significant amounts of C=C or C≡C stretching motion. For π bonded molecules, these vibrations are accompanied by periodic charge transfer to the metal. This may lead to appreciable chemical enhancement of the Raman signal according to the mechanism proposed in /132/. As ν_3 of C_2D_4 contains considerably less C=C motion than the corresponding vibration of C_2H_4, it is expected to be less intense in SER spectra (as observed). A different interpretation of the selective enhancement of some ethylene

Table 4. Vibrational frequencies of ethylene in different environments. (a): Gaseous C_2H_4, after /408/; (b) Raman study of $[Ag(C_2H_4)]^+BF_4^-$, after /418/; (c) infrared study of $(C_2H_4)Ag$ in Ar matrix, after /421/; (d) EELS study of C_2H_4 on Ag(110), T_s = 110 K, after /385/; (e) infrared reflection study of C_2H_4 on evaporated silver film, T_s = 270 K, after /451/; (f) SERS from C_2H_4 on coldly evaporated silver film, T_s = 120 K; values in parantheses have been calculated from measured C_2H_4-frequencies after /408/. After /460/

Mode Symmetry	(a) Gas		(b) Coordinated	(c)	(d) EELS Ag(110)		(e) IR Evap. Ag film	(f) SERS Coldly evap. Ag film		
	C_2H_4	C_2D_4	$[(C_2H_4)Ag]^+$	$(C_2H_4)Ag$	C_2H_4	C_2D_4	C_2H_4	C_2H_4	C_2D_4	$^{13}C_2H_4$
$\nu_{Ag-C_2H_4}$			275					182	170 (170)	180 (176)
ν_{10}, B_{2u}	826	595	873 (?)					824	603 (595)	831 (824)
ν_8, B_{2g}	940	780						939	776 (778)	932 (931)
ν_7, B_{1u}	949	720	971/990	794	967	726	985	971	732 (738)	970 (970)
ν_4, A_u	1027	726						1080	(767)	(1080)
ν_6, B_{1g}	1220	1000						1216	(995)	1191 (1207)
ν_3, A_g	1344	985	1320	1152/1132	1323	1000		1331	986 (975)	1313 (1315)
ν_{12}, B_{3u}	1444	1078						1448	(1079)	1442 (1445)
ν_2, A_g	1630	1518	1579	1476	1565	1468		1595	1477 (1485)	1557 (1553)
ν_{11}, B_{3u}	3012	2201						2961	2195 (2162)	(2958)
ν_1, A_g	3014	2262	3009		3025	2250	2850/3000	2999	2240 (2238)	2985 (2992)
ν_5, B_{1g}	3084	2315						3090	2337 (2325)	3096 (3083)
ν_9, B_{2u}	3105	2342								

modes is proposed in /67/. Here the selection rules, which govern the enhancement by photon driven charge transfer excitations, are believed to explain the observations. Yet another process, namely bonding induced changes in the intensity distribution of Raman spectra as outlined in the following sections (5.1.2 , 3), might be involved (vibrational spectra of Zeise's salt, for instance, also display only very weak C-H stretching bands /417/).

v) Remaining lines not assigned to ethylene skeletal vibrations are caused by "impurities". Some are due to isotope impurities such as C_2HD_3 (1509 cm^{-1}, ν_2), $C_2H_2D_2$

Table 5. Comparison of various SER vibrational data for ethylene on silver. (a) - (d): adsorption on coldly evaporated films; (a): "first layer signal" from bulk C_2H_4 layer deposited at T_s = 11 K, after /135/; (b): exposure of 100 L (uncorrected) at $T_s \approx$ 100 K, after /458/; (c): like (a), after /133/; (d): exposure of 36 L at T_s = 120 K, after /460/; (e): colloidal silver particles in solid C_2H_4/Ar matrix at $T_s \approx$ 10 K, "first layer signal", after /402/. Values in parentheses for (d) give relative line intensities (intensity of the strongest line has been set to 100)

Mode Symmetry	(a)	(b)	(c)	(d)	(c)	(d)	(e)
	Coldly Evaporated Ag film						Ag Colloid in matrix
	C_2H_4	C_2H_4	C_2H_4	C_2H_4	C_2D_4	C_2D_4	C_2H_4
$\nu_{Ag-C_2H_4}$				182 (?)		170 (?)	150
ν_{10}, B_{2u}	825		825	824 (0)	587	603 (0)	
ν_8, B_{2g}	955	968	955	939 (2)	761	776 (2)	957
ν_7, B_{1u}	977		977	971 (25)	718	732 (25)	
ν_4, A_u	1075	1035 (?)	1075	1080 (2)			
ν_6, B_{1g}				1216 (1)			
ν_3, A_g	1330	1333	1325	1331 (100)	972	986 (25)	1320
ν_{12}, B_{3u}				1448 (0)	1067		
ν_2, A_g	1585	1593	1585	1595 (69)	1464	1477 (100)	1586
ν_{11}, B_{3u}	2975		2975	2961 (1)	2226	2195 (1)	2969
ν_1, A_g	2996	2975 (?)	2996	2999 (2)	2242	2240 (3)	
ν_5, B_{1g} / ν_9, B_{2u}	3073		3073	3090 (1)		2337 (0)	

(1546 cm^{-1}, ν_2), and $^{12}C^{13}CH_4$ (1575 cm^{-1}, ν_2, and 1321 cm^{-1}, ν_3). Some are "characteristic impurity lines", which are frequently observed in SER spectra from silver after exposure to, for instance, CO (lines around 2110 cm^{-1} and between 1750 cm^{-1} and 1900 cm^{-1}, see Chapt. 6) or to oxygen (lines around 1065 cm^{-1} and 695 cm^{-1}, see Chapt. 7; the impurity band at 1065 cm^{-1} may obscure ν_{12} (C_2D_4) and ν_4 ($^{13}C_2H_4$), which cannot be identified in the spectra). Further weak lines are of unknown origin (bands around 330 cm^{-1}, 1130 cm^{-1}, 1365 cm^{-1}, and at 1656 cm^{-1}).

SER data are compared with vibrational frequencies of ethylene in various environments in Table 4. Assignment of the lines has been performed by comparison to gas phase values and to isotope frequencies calculated /408/ from the measured C_2H_4 SER frequencies (the calculated values are those in parentheses in the last two columns of Table 4). SER frequencies are only slightly shifted with respect to gas phase values. They agree with vibrational frequencies of ethylene on Ag(110) /385/ and of $[Ag(C_2H_4)]^+BF_4^-$ /418/ (the surprisingly small frequencies of ν_2 and ν_3 for matrix-isolated $(C_2H_4)Ag$ have been interpreted in terms of extensive coupling of these modes /421/ in this species). The largest (except ν_4) shift is observed for the C-C stretching mode ν_2 (downward by 35 cm^{-1}). This points to π bonding of the molecule to the metal. Hence, SER data are consistent with the adsorption geometry proposed in /385/: without being significantly disturbed, ethylene is weakly π bonded with its plane parallel to the surface. This lowers the symmetry from D_{2h} to C_{2v}, which may render some otherwise Raman inactive modes Raman active /135/.

Table 5 summarizes all SER data of the system ethylene/silver published so far. For adsorption on coldly evaporated films, observed spectral features agree within experimental accuracy, and mode assignment is consistent. SER data from colloidal silver particles (diameter $\approx$ 10 nm) in a solid Ar/C_2H_4 matrix /402/ are similar to those form coldly evaporated films.

Recently, SER studies have been extended to ethylene on coldly evaporated copper films /462/. Table 6 summarizes the results where only those modes are displayed which are observed in SER spectra from Cu. SER data from evaporated films and EELS data from Cu(100) /450/ agree very well. Compared to the system C_2H_4/Ag, ν_2 and ν_3 are shifted to smaller energy. This points to a stronger bond of ethylene to Cu than to Ag. The molecule is, however, still only weakly perturbed, and its adsorption geometry is presumably similar to that on silver /450/ (stronger interaction as to, for instance, Pt accompanied by considerable rehybridization shifts the frequency of ν_2 to 1230 cm^{-1} /430/). As for silver, vibrational frequencies of matrix-isolated $(C_2H_4)Cu$ disagree with those of adsorbed ethylene.

Similar to some pyridine modes, SER bands of ν_2 and ν_3 of ethylene are asymmetrically shaped (Fig. 34). The degree of asymmetry depends on the exciting frequency. It is most pronounced for red light excitation (Fig. 34, left). For isotope mixtures, the shape does not depend on the share of the constituents (Fig. 34, right). The carbon-13 labeled compound contains $\approx$ 10% of $^{13}C^{12}CH_4$. Therefore two lines are expected for ν_2 as well as for ν_3. These are almost separated for the stretching mode ν_2. The bands display the same shape as is clearly seen, when scaling them to the same peak intensity (dotted line, Fig. 34, bottom right). Possible explanations for these observations will be given below.

Table 6. Vibrational frequencies of ethylene bonded to copper. (a): Gaseous, isolated C_2H_4, after /408/; (b): infrared study of $(C_2H_4)Cu$ in argon matrix, after /419/; (c): EELS study of C_2H_4 on Cu(100), T_s = 80 K, after /450/; (d): SERS from C_2H_4 on coldly evaporated copper film, T_s = 120 K, after /462/; (e): range of C-H stretching frequencies was not accessible in this experiment. Remaining ethylene vibrations have not been observed in the SER spectra

Mode Symmetry	(a) Gas C_2H_4	(b) Coordinated $(C_2H_4)Cu$	(c) EELS Cu(100) C_2H_4	(d) SERS Cu Film C_2H_4
ν_7, B_{1u}	949	840	903	903
ν_6, B_{1g}	1220			1192
ν_3, A_g	1334	1155/1164	1290	1284
ν_2, A_g	1630	1475	1556	1550
ν_1, A_g	3014	3120	2992	(e)

Excitation profiles of SER ethylene lines from coldly evaporated silver films exhibit a resonance in the visible /281/. The resonance is centered at 2.25 eV for ν_3, and it is slightly shifted to $\approx$ 2.30 eV for ν_2 (Figs. 20 and 21). These profiles have been attributed to excitation of electromagnetic resonances in roughness features of the silver surface (see Chapt. 4). In addition, the excitation profile might be influenced by the charge transfer band of ethylene on coldly evaporated silver films at 2.8 eV /263/ (a corresponding excitation of matrix-isolated $(C_2H_4)Ag$ has been observed at 2.15 eV /422/). It is possible, that the high energy slope of the ethylene SER excitation profiles contains contributions from a weak resonance Raman effect due to the charge transfer excitation.

Coverage Dependence. Figure 35 displays the development of the SER signal of ν_3 with exposure /459/. A dose of 10^{-3} L, leading to at most one thousandth of a monolayer coverage, results in an easily detectable signal of $\approx$200 cts/s peak intensity (because of experimental limitations we were not able to realize smaller exposures). This sensitivity is superior to that obtained with other techniques like IRAS /17/ and EELS /16,463/. The peak intensity increases with exposure to $\approx 5 \cdot 10^3$ cts/s for 3.6 L and drops slowly after further exposure. The characteristic asymmetrical shape of the SER line is most pronounced for small exposures. With increasing dose, the line shifts to greater energy (by $\approx$ 5 cm^{-1}, Fig. 36b), narrows, and becomes more

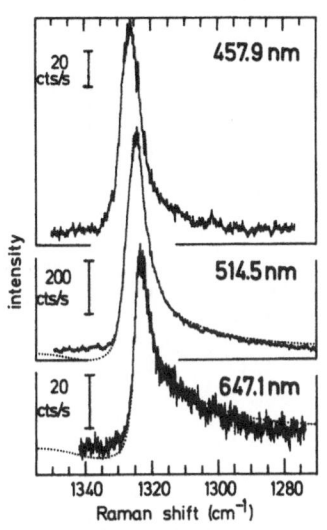

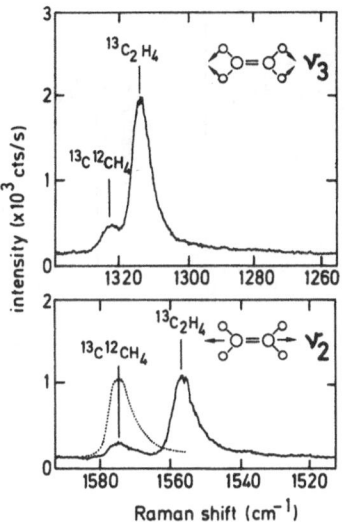

Fig. 34

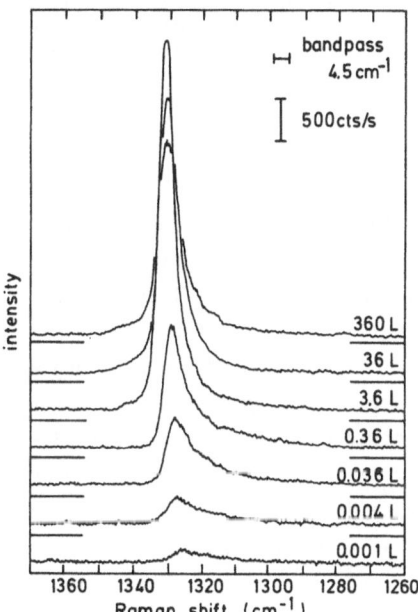

Fig. 34. Influence of excitation frequency on line shape of ν_3 (Ag film exposed to 36 L of C_2H_4, left), and SER features from Ag film exposed to 36 L of a 1:9 mixture of $^{13}C^{12}CH_4$ and $^{13}C_2H_4$ (right). All spectra have been taken with 4.5 cm^{-1} bandpass; 60 mW of 457.9 nm, 200 mW of 514.5 nm, and 190 mW of 647.1 nm radiation. After /460/. Dotted lines have been calculated assuming a Fano resonance (for details see /67/)

Fig. 35. Development of SER signal from scissors mode of ethylene (ν_3) on Ag film with exposure (excitation with 200 mW of 514.5 nm radiation). After /459/

symmetrical [Fig. 36c; note, that the full width at half maximum contains a contribution from the finite bandpass of the spectrometer; the true line width of ν_3, measured with small bandpass, is 4.3 cm^{-1} (3.6 L exposure)].

The integrated line intensity of ν_3 as a function of exposure is shown in Fig. 36a. Plotting exposure on a logarithmic scale, the intensity increases almost linearly, with a change of the slope at ≈ 1 L. For exposures greater than $\approx 10 - 50$ L,

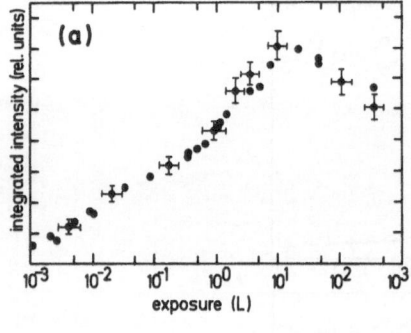

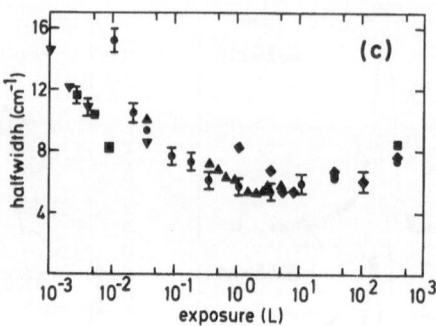

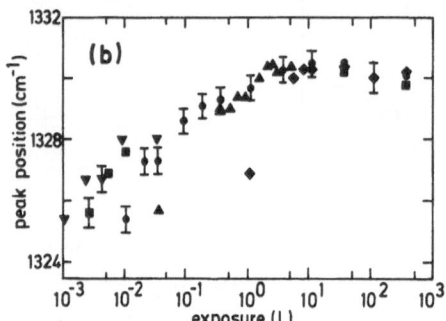

Fig. 36. Integrated intensity (a), spectral position (b), and halfwidth FWHM (c) of ν_3 SER line from C_2H_4 on Ag film as a function of exposure. Different symbols represent different experimental runs (b and c). FWHM-values (c) contain contribution from finite bandpass (4.5 cm^{-1}) of the spectrometer. After /459,460/

a decrease of the SER intensity of ν_3 is observed. The C-C stretching mode ν_2 behaves similarly /459/. The decrease might be explained with "poisoning" of the surface. Impurities in the used ethylene gas or formed on the surface of the coldly evaporated film might displace the weakly bonded ethylene and contaminate the silver surface. Evidence for such effects has been found after extensive exposure of coldly evaporated Ag films with CO, where the development of a line at ≈ 2110 cm^{-1} goes along with a decrease of the characteristic CO peaks (see Chapt. 6). A quantitative interpretation of the almost linear increase of the SER intensity ($\lesssim 10$ L) is difficult. Because of the inhomogeneity of the surface of coldly evaporated silver films, adsorption is not expected to follow a simple law. The data displayed in Fig. 36a presumably represent a superposition of various adsorption isotherms each characteristic for a special site (C_2H_4 SER spectra do not show *drastic* effects due to adsorption on different sites, multiple lines for a given vibration are not observed; the characteristic asymmetrical line shape might, however, reflect the inhomogeneity of the surface).

Annealing Behaviour. SER lines from ethylene on coldly evaporated silver films change characteristically upon annealing (Fig. 37, scissors mode ν_3). Starting at ≈ 160 K, the line broadens, becomes more asymmetrical, and the peak position shifts to smaller energy (Figs. 37a and 37b). Simultaneously, the integrated line intensity decreases (Fig. 37c). The variation of the spectral features with *increasing* temper-

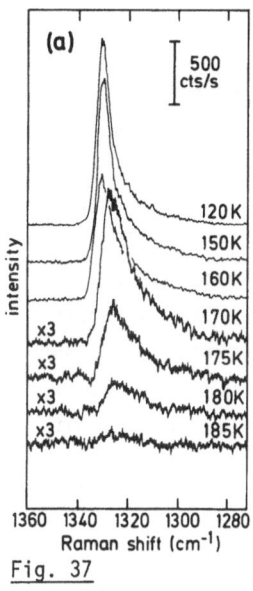

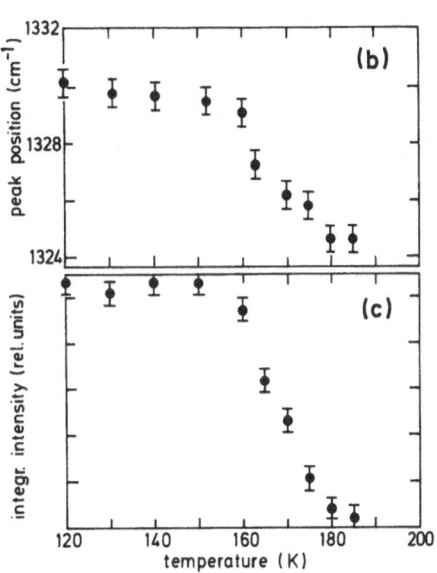

Fig. 37

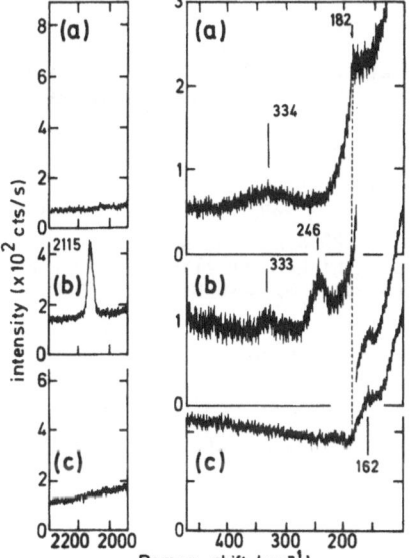

Fig. 37. Annealing of SER signal from ethylene (3.6 L) on coldly evaporated silver film (scissors mode ν_3; $\Delta T \approx 1$ K/min). (a): Change of spectrum (200 mW of 514.5 nm radiation, 4.5 cm^{-1} bandpass); (b), (c): development of peak position and integrated intensity. After /460/

Fig. 38. Annealing of SER features from coldly evaporated Ag film exposed to 3.6 L of ethylene. (a): spectral features at 120 K; (b): sample annealed to 190 K, recooled to 120 K, measured; (c): sample annealed to 250 K, recooled to 120 K, measured. 200 mW of 514.5 nm radiation, 4.5 cm^{-1} bandpass. After /460/

ature between 160 K and 185 K is similar to that of corresponding features with *decreasing* exposure (Fig. 35). Hence, the data of Fig. 37 reflect the changes of ethylene coverage and suggest desorption of ethylene between 160 K and 185 K. For C_2H_4 on Ag(110), desorption has been observed at $\approx$ 170 K /385/ in good agreement with our result.

After desorption of ethylene from coldly evaporated silver films, *new* lines are observed in SER spectra at 246 cm^{-1}, 333 cm^{-1}, and 2115 cm^{-1} (Fig. 38). No other bands are detected. Note that the shoulder at 182 cm^{-1} has disappeared which corroborates the assignment to $\nu_{Ag-C_2H_4}$ given above. Only a shoulder at 162 cm^{-1} is left, which has previously been attributed to disorder induced Raman scattering from bulk silver phonons (see Chapt. 3). As will be outlined later, the features displayed in Fig. 38b are tentatively attributed to an adsorbed bidentate acetylide species. It is assumed, that part of the adsorbed ethylene decomposes into this species upon annealing (details unknown). After further annealing of the sample to ≈ 250 K, all SER features have disappeared (Fig. 38c), which might reflect the decreasing SERS activity of the silver surface in this temperatures range (see preceeding chapter).

When coldly evaporated silver films are warmed to temperatures *above* the desorption temperature of ethylene, subsequently are re-cooled to 120 K, and re-exposed to C_2H_4, spectral changes with annealing temperature (Fig. 39) similar to those with

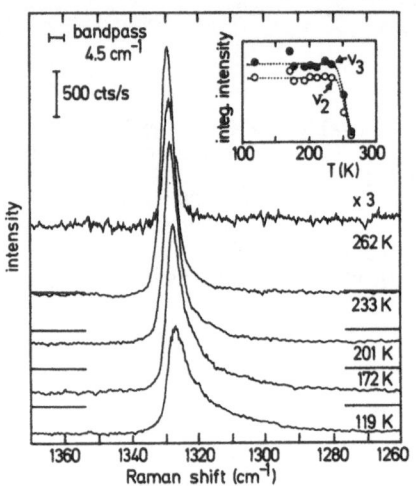

Fig. 39. Effect of *substrate* annealing on SER signal from scissors mode of C_2H_4 on Ag film. Exposed (1 L), coldly evaporated film has been warmed to the temperature indicated (ΔT ≈ 1 K/min), re-cooled to 120 K, and re-exposed to 1 L of C_2H_4; then a SER spectrum was taken at 120 K and the same procedure repeated with higher annealing temperature, etc. (note that C_2H_4 desorbs at ≈ 170 K). Inset shows development of integrated line intensity. After /460/

coverage (Fig. 35) are observed. This holds for the line shape and position, but not for the intensity. The integrated line intensity does not change up to annealing temperatures of ≈ 230 K (see inset in Fig. 39). When warming the SERS active film to T ≳ 230 K, SER intensities drop rapidly. This indicates the loss of SERS activity of the film due to corresponding structural changes of the silver surface. We recall, that the drop of pyridine SER intensities for T ≳ 220 K (Figs. 17 and 18) has been explained in this way in Chapt. 4. In closing we emphasize, that the spectral changes displayed in Fig. 39 are caused by changes of the silver surface topography, whereas the similar development (concerning line shape and position) shown in Fig. 35 seems to be an exclusively coverage dependent effect.

General discussion. SER data from adsorbed ethylene point to weak interaction of the molecule with silver. Compared to gas phase values, vibrational frequencies are only slightly shifted in agreement with EELS results from single crystal planes /385/. The largest shift upon adsorption is observed for the C-C stretching mode ν_2. Therefore, adsorption of the molecule with its plane parallel to the surface by forming a weak π bond as suggested in /385/ seems to be reasonable. Bonding to Cu is stronger as indicated by the more pronounced shift of the vibration frequencies. For C_2H_4 on Ag some modes give extremely strong SER signals: an exposure of 10^{-3} L leads to an easily detectable signal.

Like some other SER lines of adsorbed molecules (see Chapts. 4,6), the strongest ethylene lines are asymmetrically shaped. One might think of three explanations:

i) Dipole-dipole interaction between vibrating molecules (dynamical dipole moment vertical) in a disordered, incomplete layer on the surface leads to line broadening, shift to smaller energy, and increasing asymmetry of the vibrational line shape with decreasing coverage /464/.

ii) The interference of a discrete state with a continuum of states may result in characteristic asymmetrical line shapes (Fano resonance, /465/). Within this frame the vibrational level could be the discrete state, which interacts with the continuum of eh-pair excitations /67/.

iii) The presence of various adsorption sites (e.g. defect sites) on the surface of coldly evaporated silver films may lead to inhomogeneous line broadening.

Although the development of the line with exposure (Fig. 35) points to (i), this model can be ruled out because of the annealing behaviour (Fig. 39) and the result for the iotope mixture (Fig. 34, right). The concept of the Fano resonance has been discussed in some detail in /67/. It easily explains the changes of the line shape with exciting frequency (Fig. 34, left). However, we never observed the intensity dip characteristic for a Fano profile (the calculated curves in Fig. 34 display it at the high energy side of the maximum). Therefore we favour (iii). Tentatively, we suggest the following interpretation scheme.

As shown recently, ethylene adsorption on Ag(110) is strongly promoted by pre-adsorbed atomic oxygen /385/. It was concluded, that the molecule bonds selectively to those silver atoms, on which a positive charge had been induced by the oxygen atoms. Accepting these ideas, we assume /459/, that C_2H_4 adsorbs on coldly evaporated silver films at 120 K only on special, positively charged sites as, e.g., adatoms at steps, kinks, steps itself /377/, as well as on certain "impurity" sites (details unknown). In general, bond strength and hence vibrational frequencies depend on the adsorption site. For adsorbed C_2H_4, ν_2 as well as ν_3 shift to smaller energy with increasing bond strength /433/. We expect strongest interaction (bonding) at the most positively charged defect sites available (here step adatoms, see Fig. 31). It is reasonable to assume, that these sites will preferentially be occupied first upon exposure. More weakly bonded molecules (e.g. on step sites) will, however, constitute the majority species for saturation coverage because of the distribution of the various defect sites on the surface of coldly evaporated silver films (note, that the sites, which offer stronger bonding, are the more scarce ones; therefore, depending on coverage, more or less asymmetrical line shapes are expected).

Within this frame, the faster increase of the SER intensities with exposure above $\approx 1\,L$ (Fig. 36a) is due to the growing density of positively charged adsorption sites (caused by as yet unknown "impurity" adsorbates /385,459/). The same phenomenon leads to the decrease of the SER intensity for extensive exposure ($\gtrsim 10 - 50\,L$) because of the "poisoning" mentioned above.

The variation of the line shape with exposure (Fig. 35) and upon annealing (Fig. 37a) mirrors changes in the *distribution* of the ethylene molecules on the various adsorption sites. Preferential occupation of sites offering stronger bonding for small exposure is responsible for the, correspondingly, broad SER feature, whereas the band is dominated by the more weakly bonded majority species for saturation exposure. Hence it seems to narrow and to shift with exposure (Fig. 35), which is actually due to the fact that intensity is added only on its high energy side. Upon annealing, the most weakly bonded species will desorb first. A sensitive, temperature programmed desorption study should allow to separate differently bonded species. Because of experimental limitations, corresponding investigations have not been performed. However, desorption of the most weakly bonded species at $T \lesssim 170\,K$ could clearly be separated (Fig. 40, $\Delta T \approx 1\,K/min$).

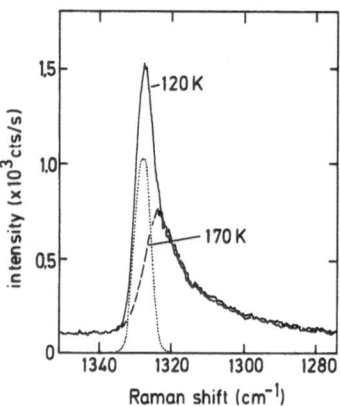

Fig. 40. Annealing of scissors mode of C_2H_4 in SER spectrum from coldly evaporated Ag film (3.6 L exposure). Solid line: band measured at 120 K; dashed line: band measured after annealing to 170 K ($\approx 1\,K/min$). Dotted line gives difference of the two spectra. Spectra were recorded with 200 mW of 514.5 nm radiation and 4.5 cm^{-1} bandpass. After /460/

It is clear, that annealing affects also the surface topography of the silver film. The most outstanding defect sites offering the strongest bonding will anneal first. Hence, spectral features of C_2H_4 on annealed films should gradually assume the shape expected for the most weakly bonded majority species with increasing annealing temperature, which is indeed observed (Fig. 39). Note that this interpretation implies a chemical enhancement mechanism rather independent of the detailed structure of the adsorption site, since the integrated SER intensity does *not* change over a wide range of annealing temperatures.

Every component of isotope mixtures is distributed on the various adsorption sites in the same way, independent of its fraction. Consequently, line shapes should be identical, which is observed (Fig. 34, right). The sensitivity of the line shape to the frequency of the exciting radiation (Fig. 34, left) can be explained by recalling the electromagnetic enhancement mechanism due to surface plasmon resonances in surface bumps (see Chapt. 2). The resonance of a high density of small bumps is red shifted with respect to that of a lower density of somewhat larger bumps. Therefore red light excites most efficiently electromagnetic resonances on parts of the surface with a relative high density of defect sites (because of the large curvature of small bumps), whereas blue light probes smoother parts. Consequently, SER lines from adsorbed C_2H_4 should assume a more symmetrical shape and shift to somewhat larger energy with decreasing excitation wavelength (as observed, Fig. 34, left).

We think, that asymmetrical SER line shapes of some other adsorbates on Ag (pyridine, Chapt. 4, or carbonaceous deposits, Chapt. 6) have to be interpreted as outlined in the preceeding paragraph for C_2H_4. An alternative explanation for the asymmetrical line shape and the mode selective enhancement for C_2H_4 on silver based on the photon driven charge transfer model is suggested in /67/.

Although the proposed model explains C_2H_4 SER features quite suggestively, it contains some degree of speculation. Further experiments are necessary to consolidate the conclusions. A natural and interesting extension is to deliberately pre-dosed surfaces as, for instance, oxygen pre-dosed silver surfaces, which are much more reactive than clean surfaces (e.g. /466/).

Finally, we point again to the drawback from which interpretations of SER results generally suffer at present. As long as the nature of SERS is not understood in detail, it cannot be excluded with certainty that observed trends of SER features mirror a property of the SER process itself. A coverage dependence of the enhancement, which has been neglected in the discussion outlined above, but is not entirely inconceivable, would further complicate the interpretation of the experimental results.

5.1.3 Propylene and Butylene

Besides C_2-alkenes, propylene and some butylenes are the only open-chain hydrocarbons, which so far have been studied by SERS. Propylene has 21 vibrational modes (14A' + 7A", point group C_s), which are all Raman active. 1-butene belongs to the same symmetry group (C_s), whereas isobutylene and cis-2-butene have C_{2v} and trans-2-butene C_{2h} symmetry. For the butenes with C_s and C_{2v} symmetry all modes are Raman active (1-butene: 19A' + 11A"; isobutylene and cis-2-butene: $10A_1 + 5A_2 + 9B_1 + 6B_2$). Only the A_g and B_g modes are Raman active for trans-2-butene ($10A_g + 6A_u + 5B_g + 9B_u$). Geometrical and vibrational data of the hydrocarbons are summarized in /1,467/, a detailed

Table 7. SER vibrational frequencies of some C_3- and C_4- open-chain hydrocarbons on Ag compared to gas phase values and data of some complexes. (a) , (d) , (g): gas phase data after /467/, /473/, and /472/ respectively; (e) , (f): vibrational frequencies of [Ag(trans-2-butene)]$^+BF_4^-$/[Ag(cis-2-butene)]$^+BF_4^-$ /418/ and [Ag(trans-2-butene)]$^+NO_3^-$ / [Ag(cis-2-butene)]$^+NO_3^-$ /474/; (b) , (c): SER data after /135/ and /401/ respectively. Only the modes detected in SER spectra have been considered, and an asterisk means assignment performed by the present author

Propylene			Isobutylene			1-Butene			Trans-2-Butene					Cis-2-Butene				
Mode Symmetry	(a) Gas	(b) SERS	Mode Symmetry	(a) Gas	(c) SERS*	Mode Symmetry	(a) Gas	(c) SERS*	Mode Symmetry	(d) Gas	(e) Complex*	(f) Complex*	(c) SERS	Mode Symmetry	(g) Gas	(e) Complex*	(f) Complex*	(c) SERS
ν_{14},A'	428	440	ν_{10},A_1	378	380	ν_{18},A'	437	430	ν_{30},B_u	260			262	ν_{10},A_1	302			300
ν_{20},A''	578	610	ν_{29},B_2	431	438	ν_{27},A''	788	790	ν_{10},A_g	501	500		501	ν_{15},A_2	394		434	397
ν_{13},A'	920	925	ν_{24},B_1	802	815	ν_{17},A'	853	852	ν_{14},B_g	755			756	ν_{30},B_2	687			711
ν_{17},A''	1045	1050	ν_9,A_1		824	ν_{26},A''	912	920	ν_9,A_u	861	871		865	ν_9,A_1	874	869	876	864
ν_{11},A'	1171	1170	?		838	ν_{24},A''	1020	1020	ν_{19},A_u	973	980		977	ν_{19},B_1	974	971		978
ν_{10},A'	1297	1300	?	887	891	ν_{23},A''	1264	1260	ν_{13},B_g	1041			1037	ν_8,A_1	1010	1013		1008
ν_9,A'	1378	1380	ν_{28},B_2	1053	1060	ν_{13},A'	1294	1296	ν_{18},A_u	1042	1042		1088	ν_{13},A_2	1038	1041		1034
ν_8,A'	1420	1420	ν_8,A_1	1379	1386	ν_{10},A'	1420	1415	ν_7,A_g	1304	1302	1290	1302	ν_7,A_1	1257	1250	1256	1255
ν_6,A'	1650	1612	ν_7,A_1			ν_9,A'	1450	1435	ν_{27},B_u	1364			1357	ν_{27},B_2	1357	1360		1353
ν_5,A'	2924	~2950	ν_{20},B_1			ν_8,A'	1457	1458	ν_6,A_g	1386	1390		1377	ν_6,A_1	1385	1387	1382	1375
ν_{15},A''	2956		ν_6,A_1	1412	1417	ν_{22},A''			ν_5,A_g	1441			1436	ν_{25},B_2	1445	1441		1440
ν_5,A'	2956		ν_4,A_1	1660	1602	ν_7,A'	1645	1600	ν_{12},B_g	1452	1457	1450	1447	ν_5,A_1	1462	1451		1455
									ν_4,A_g	1668	1615	1612	1632	ν_4,A_1	1662	1597	1598	1621

discussion of vibrational properties may be found in /468,469/ (propylene), /470, 471/ (isobutylene), /472/ (cis-2-butene), and /473/ (trans-2-butene). Metal-olefin complexes have frequently been investigated /474 - 478/. Vibrational data of coordination compounds with silver /418,479,480/ are especially interesting for our purposes. Like other hydrocarbons, propylene and the butenes are transparent in the visible, the lowest lying optically allowed transitions are found in the UV /424,481,482/. Hence only ordinary Raman scattering is observed from these molecules.

Only few investigations have been dedicated to absorption of C_3- and C_4-hydrocarbons on metal surfaces (e.g. /441,483 - 486/), although propylene and butene oxidation over, for instance, copper catalysts /487,488/ is a process of great industrial importance. To the knowledge of the author, vibrational studies of propylene or butene on metals from group Ib have not been performed so far (with the exception of SER studies).

For the SER vibrational studies, thick layers of propylene /135/ and some butenes /401/ (about 200 monolayers) have been condensed on silver films evaporated at 11 K. SER lines from the molecules directly attached to the metal (first monolayer) have been extracted from the spectra by comparison to ordinary spectra from the same, thick layer on oxide-covered aluminum surfaces. Table 7 summarizes the results. SER vibrational frequencies do not differ markedly from gas phase data, except for modes which contain substantial double bond stretching (ν_4 for the butenes except 1-butene (ν_7), ν_6 for propylene). The observed downward shift (40 - 60 cm^{-1}) of the C=C stretching vibrations implies that the molecules are weakly π bonded to the silver through the double bond /401/. This conclusion is corroborated by comparison to silver coordinated C_3- and C_4-olefins, whose C=C stretch also experiences a downward shift caused by the π bond to the silver atom /418,479,480/ (this shift is slightly stronger than in the SER case).

Like other SER spectra, the spectrum of trans-2-butene contains some normally Raman silent bands with appreciable intensity (/401/; e.g. ν_{19}, a mode belonging to the A_u representation which is neither infrared nor Raman active). Mode selective enhancement is observed: the modes containing substantial C=C stretching character are the most prominent in SER spectra /401/. By contrast, C-H stretching vibrations are only weakly pronounced, similar to pyridine and C_2-hydrocarbon SER spectra. The mode selective enhancement might be explained as outlined in Sect. 5.1.2 for ethylene. Again it is interesting to note, that the intensity of C-H stretching bands is considerably weakened upon coordination of alkenes to metals /417,418,479/. This has been assigned to the decrease of the π electron density at the trigonal carbon atoms in the olefin complexes /479,489/. Similar effects may be important for other adsorbed olefins.

In summary, SER vibrational data from propylene and some butenes on silver at 11 K point to weak bonding to the metal via the π electrons of the C=C double bond.

5.1.4 Acetylene

Acetylene is a linear and symmetrical molecule (point group $D_{\infty h}$). It has five normal modes of vibration: the three "gerade" modes are Raman active $(2\Sigma_g^+ + 1\Pi_g)$, the two "ungerade" modes are infrared active $(1\Sigma_u^+ + 1\Pi_u)$. Its geometrical and vibrational structure has extensively been studied /1/, vibrational frequencies of gaseous and solid C_2H_2 as well as of the deuterated species are well known /490 - 493/. Alkynes are known to form both, σ bonded organometallic compounds as well as π bonded co-ordination complexes /476 - 478,494/. Copper and silver compounds of acetylene (generally substituted) have been investigated (e.g. /495 - 497/), vibrational data of some complexes /498,499/ as well as of $Cu(C_2H_2)_n$ (n = 1,2) isolated in an Ar matrix at liquid He temperature /500/ have been published. With respect to the uncoordinated acetylenes, the C≡C stretching frequency usually shifts to lower frequencies upon complex formation indicating reduction of the acetylenic bond order (shifts of 270 - 300 cm^{-1} (Cu) and 150 - 200 cm^{-1} (Ag) have been measured /498/). Many complexes are unstable and sensitive to light /478,499/. Electronic absorption spectra of isolated acetylene /424,427,501/ start with a weakly allowed /502/ transition peaking at 6.35 eV followed by the strong first Rydberg transition at 8.16 eV. Electron impact spectra /425,428/ reveal at least two additional singlet-triplet transitions at 5.2 eV and 6.1 eV respectively /503/. Matrix-isolated copper-acetylene complexes display an absorption band in the visible around 450 nm /500/.

Acetylene adsorption on metals has been the subject of numerous studies. When adsorbed at low temperature on the transition metals of group VIII, the molecule is strongly distorted. Considerable rehybridization accompanied by reduction of the acetylenic bond order to $\approx 1 - 2$ has been observed for low index Ni single crystal faces /431,433,504,505/, for Pt(100) /506/ and Pt(111) /430,431,433,507,508/, and for Pd(111) /431,440,509/, Rh(111) /434/, and Fe(110) /435/. Qualitatively similar behaviour has been reported for adsorption on Ir(100) /510/ and Ru(0001) /511/. Acetylene interacts also strongly with Re /447,448/ and with W /512/. Similar to ethylene, new species are usually formed when warming exposed surfaces to room and higher temperature (or upon adsorption at temperatures $\gtrsim$ 300 K; Ni: /504,513,514/; Pt: /430, 442,443/; Rh: /434/; Fe: /435/; W: /512/; Pd: /440/). The strong interaction of acetylene with stepped Ni(111) surfaces /436,515/ is particularly interesting. On these surfaces, acetylene instantaneously dehydrogenates to C_2 which further decomposes into carbon atoms, even at 150 K. Strong interaction of acetylene has also been reported with silica-supported metals (e.g. /516/: Ni and Pt) and evaporated films (e.g. /446/: Ni and Pd).

Only a few papers treat the adsorption of acetylene on the noble metals of group Ib /517-522/. Contradicting results have been published for Cu(111). Whereas UPS data suggest weak interaction with little perturbation of the molecule /452/, Auger spectra seem to indicate strong interaction of possibly di-σ, π type accompanied by rehybridization to sp^2 - sp^3 /432/. Early IRTS data from C_2H_2 on fine copper particles formed by an exploding wire /518/ agree with the latter, data from supported copper /519/, however, do not. Adsorption of acetylene on Ag(110) has recently been studied in some detail by temperature programmed reaction spectroscopy, photoemission spectroscopy, AES, LEED, EELS, and titration reactions /520 - 522/. On clean Ag(110), C_2H_2 adsorbs at 100 K without rehybridization with its C≡C axis parallel to the surface and desorbs without reaction between 100 K and 160 K /522/. Oxygen pre-dosed surfaces are much more reactive. C_2H_2 reacts with oxygen *atoms* on Ag(110) to form adsorbed monodentate acetylide species C_2H and water. At 270 K, C_2H disproportionates to yield bidentate acetylide C_2 (adsorbed) and acetylene (which desorbs) /521/. UPS and EELS data suggest a C-C bond order of approximately three in C_2H. It is proposed, that the monodentate acetylide resides on Ag ridge atoms and bridges the troughs of the surface, where weak π interactions pull the C≡C bond parallel to the surface /522/. Similarly, adsorbed C_2 shows no evidence for rehybridization /521/.

General SER Features. Figure 41 displays Raman spectra from coldly evaporated silver films exposed to 36 L of various acetylene isotopes as indicated /523/. A ten times smaller dose gave essentially the same spectra. Therefore the curves of Fig. 41 correspond to saturation coverage. As multilayers are not stable under our experimental conditions, the Raman signal is due to at most a monolayer of adsorbed molecules and hence enhanced. The number of observed peaks is surprisingly high. The most intense features are compsed of two peaks, e.g. 1934/1888 cm^{-1}, 789/756 cm^{-1}, and 673/635 cm^{-1}

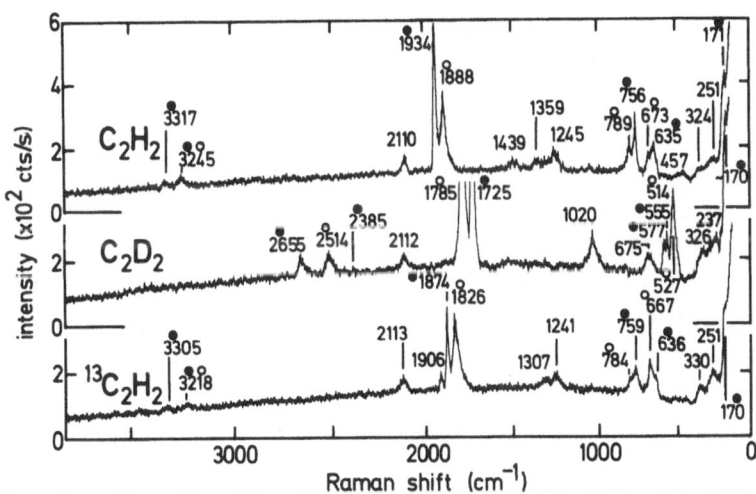

Fig. 41. SER spectra from coldly evaporated silver films exposed to 36 L of acetylene. Filled-in circles mark lines from adsorbed acetylene, open circles those from monodentate acetylide C_2H (see text). 200 mW of 514.5 nm radiation, 4.5 cm^{-1} bandpass. After /523/

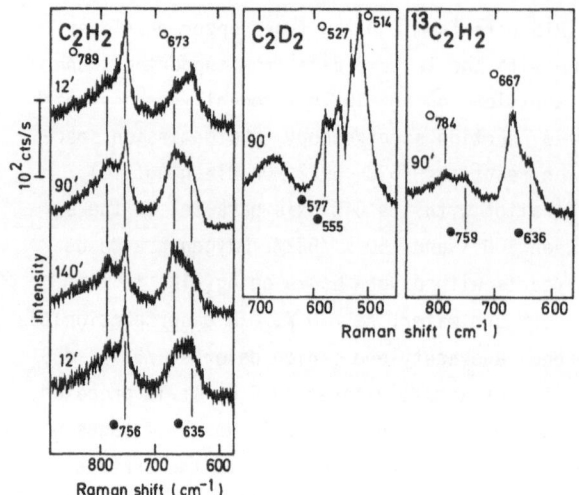

Fig. 42. Expanded bending mode region of spectra displayed in Fig. 41. Series of C_2H_2 spectra shows illumination induced spectral changes (value to the left of each spectrum is the illumination period in min; bottom spectrum has been taken after focusing on a new spot on the surface). After /525/

for C_2H_2 (the bending mode region is shown in somewhat more detail in Fig. 42). The C-H stretching vibrations (v_1, v_3), which are quite strong in ordinary Raman (IR) spectra /1/, give rise to only weak features in SER spectra (e.g. 3245/3317 cm^{-1} for C_2H_2). The corresponding lines of the deuterated species are stronger (2385/ 2656 cm^{-1}). This is similar to SER spectra of other hydrocarbons (see preceeding sections) and of pyridine (see Chapt. 4). Several peaks, marked with filled-in circles in Figs. 41, 42, disappear when warming the sample to $\approx$ 145 K. This suggests desorption of the species responsible for these lines (see also below). The same bands are also sensitive to the incident, exciting light (Fig. 42). Their intensity decreases with illumination time. Photodecomposition or laser stimulated desorption of the adsorbed species might explain this observation.

An assignment of the SER features of this weakly bonded species is given in Table 8. Vibrational energies are very similar to corresponding gas phase data /1/. This points to little perturbation of the molecule upon adsorption. Therefore SER line frequencies of C_2D_2 and $^{13}C_2H_2$ may be calculated from the measured C_2H_2 values by using the well known isotope shifts derived for the isolated molecule /524/. Calculated frequencies agree with the experimental values quite well (Table 8). There remains some uncertainty concerning the assignment of the bending modes (v_4, v_5), which has been slightly changed with respect to /523/. The silver-acetylene stretching frequency is *not* hidden in the broad feature between 220 cm^{-1} and 350 cm^{-1}. These frequencies are too high for a species already desorbing at $\approx$ 145 K. Based on a detailed study of the low frequency region presented elsewhere /525/, we assign the weak feature at $\approx$ 170 cm^{-1} to $v_{Ag-C_2H_2}$. SER frequencies agree with data of a recent EELS study of acetylene adsorption on Ag(110) /522/. The results of the latter suggest, that C_2H_2 adsorbs approximately parallel to the surface by weak π inter-

Table 8. Vibrational frequencies of acetylene in different environment.
(a): gaseous acetylene, after /1/; (b): EELS from C_2H_2 on Ag(110), T = 100 K, after /522/; (c): EELS from C_2H_2/C_2D_2 on Pt(111), T = 150 K, after /430/; (d): SERS from coldly evaporated silver film, T = 120 K; values in parentheses have been calculated from measured C_2H_2 frequencies after /524/; after/525/; (e): as (d), but T = 11 K, after /133/; (f): SERS from colloidal Ag particles in Ar/C_2H_2 matrix, T = 11 K, after /402/

Mode Symmetry	(a) Gas C_2H_2	(a) Gas C_2D_2	(b) EELS Ag(110) C_2H_2	(c) EELS Pt(111) C_2H_2	(c) EELS Pt(111) C_2D_2	(d) SERS Coldly evap. Ag film C_2H_2	(d) C_2D_2	(d) $^{13}C_2H_2$	(e) C_2H_2	(e) C_2D_2	(f) Ag Colloid C_2H_2
$\nu_{Me-C_2H_2}$			300	470	440	171	~170 (165)	~170 (165)			170
ν_4, Π_g	612	505	(650)			635	555 (531)	636 (651)	634	508	632
				985	730						
ν_5, Π_u	729	539	770			756	577 (555)	759 (785)	781	562	769
										1716	
ν_2, Σ_g^+	1974	1762		1310	1260	1934	1725 (1715)	1874 (1873)	1923	1736	1934
									1954	1778	
ν_3, Σ_u^+	3287	2427	3270			3245	2385 (2381)	3218 (3235)			
				3010	2245						
ν_1, Σ_g^+	3374	2701				3317	2655 (2645)	3305 (3286)		2645	

action and is negligibly perturbed from its gas phase state. A similar adsorption geometry is assumed for the weakly bonded species on the silver films. For comparison, Table 8 contains also vibrational data from an EELS study of acetylene on Pt(111) /430/, where strong interaction leads to considerable distortion of the molecule.

The line frequencies of the weakly bonded species are similar to SER data from acetylene condensed on silver films at 11 K /133/ and from colloidal silver particles formed by gas aggregation and isolated at low temperature in a solid Ar/C_2H_2 matrix /402/ (Table 8). All spectra exhibit normally Raman-forbidden lines, and features due to C-H stretching modes are either absent /133,402/ or only weakly pronounced (Fig. 41), which both are quite common phenomena in SERS from solid/gas interfaces.

A second set of SER lines, marked by an open circle in Figs. 41, 42, is lost at considerably higher temperature (between ≈ 170 K and ≈ 220 K). These lines are listed in Table 9. They have been attributed to more strongly bonded acetylene on "special, active defect sites" in /523/. If also the molecules on these sites essentially preserve their identity, it is, however, difficult to understand, why the C-C stretching

Table 9. Vibrational frequencies of acetylide species. (a): gaseous C_2H_2, after /1/; (b): IR spectrum of sodiumacetylide, after /526/; (c): EELS from oxygen pre-dosed Ag(110) exposed to C_2H_2, $T \lesssim 240$ K, after /522/; (d): SERS from coldly evaporated silver film, T = 120 K, after /525/; (e): IR (Raman) of methylacetylene and polymerized methylethynylsilver, after /527/

Mode	(a) Gas	(b) Coordinated	(c) EELS	(d) SERS			(e)	
			Ag(110)	Coldly evap. Ag film				
	C_2H_2	NaC_2H	C_2H	C_2H	C_2D	$^{13}C_2H$	CH_3-C_2H	$(CH_3-C_2Ag)_x$
ν_{Ag-C_2H}			300	~270	~270			
δ_{CC}	612	647	690	673	514	667	336	364
δ_{CH}	729			789	527	784	643	
ν_{CC}	1974	1867		1888	1785	1826	2142	2062
ν_{CH}	3287 3374	3225	3250	(3245)	2514	(3218)	3305	

frequency (ν_2) shifts upon deuteration by only half the value (103 cm^{-1}) observed for the weakly bonded species (209 cm^{-1}). We suggest /525/ to attribute these lines to monodentate acetylide C_2H reactively formed on "special sites" (formation of this species on Ag(110) by reaction with oxygen atoms has been reported /522/ as mentioned earlier). The interpretation is supported by vibrational data of acetylene compounds with similar bonding (Table 9). SER values agree rather well with those of sodium-acetylide /526/, and the downward shift of ν_{C-C} of C_2H compared to gaseous C_2H_2 (86 cm^{-1}) agrees excellently with the equivalent shift of polymerized methylethynyl-silver compared to its parent molecule, methylacetylene (80 cm^{-1}; Table 9). The metal-acetylide stretching mode is observed at ≈ 270 cm^{-1} as shown in somewhat more detail below. We note, that our interpretation requires either oxygen atoms /522/ or some other sort of "special sites" on the surface, which are able to partly dehydrogenize acetylene. This point as well as possible adsorption geometries are discussed below.

Many of the remaining bands in acetylene SER spectra (Fig. 41) can also be assigned. The weak feature at 1906 cm^{-1} ($^{13}C_2H_2$ spectrum) is due to ν_2 of a $\approx 10\%$ impurity of $^{13}C^{12}CH_2$ in the supply. The lines at ≈ 2110 cm^{-1} and at $\approx 250/330$ cm^{-1} are particularly interesting: they are attributed to the C-C stretching and the Ag-C_2 stretching modes of a completely dehydrogenated acetylene species C_2 on the surface (see below). They frequently accompany hydrocarbon SER spectra and are the prominent features in spectra from silver films which have excessively been exposed to CO (for a detailed discussion see Chapt. 6). The origin of the other lines is unknown, bands between 1200 cm^{-1} and 1500 cm^{-1} are presumably caused by carbonate- or graphitic carbon-deposits (see Chapt. 6).

Coverage Dependence. Figure 43 displays SER features in the region of the C-C stretching modes from coldly evaporated silver films exposed to various amounts of C_2H_2 as indicated. ν_2 of the weakly bonded species shifts from 1924 cm^{-1} to ≈1935 cm^{-1} with exposure. The line is inhomogeneously broadened. Upon annealing, intensity is first removed from the high energy side of the band (at ≈ 135 K), and the peak of the band shifts to smaller energy with increasing temperature (between ≈ 135 K and ≈ 145 K /525/). A similar behaviour has been observed for some ethylene SER lines (Sect. 5.1.2). This suggests a common explanation. The interpretation, which is based on the inhomogeneity of our silver surface, has already been outlined at the end of Sect. 5.1.2. The development of the ν_{C-C} band of the monodentate acetylide species is consistent with this interpretation: whereas the smallest exposure leads to only one line at ≈ 1835 cm^{-1}, three lines can be distinguished for greater exposure

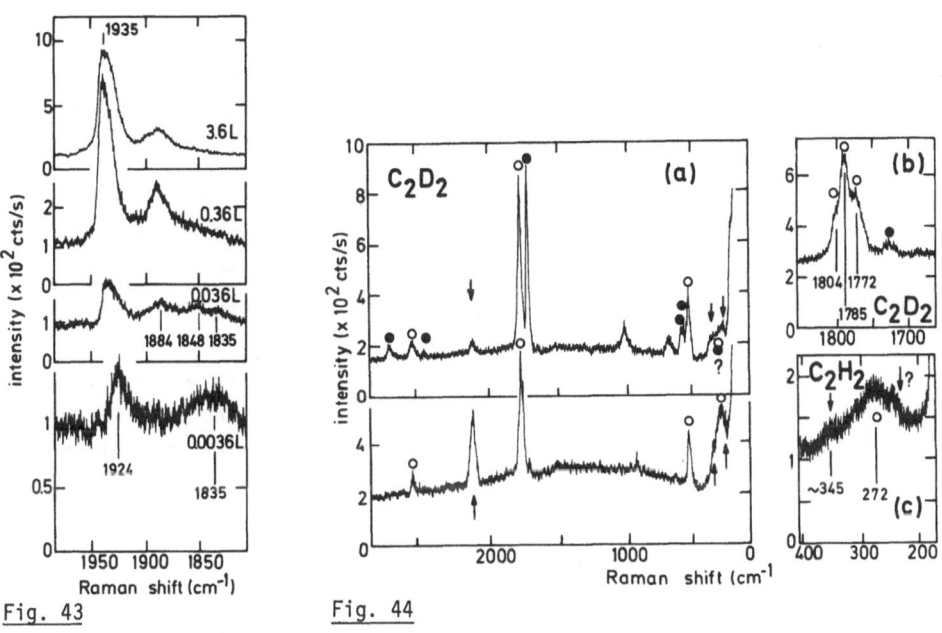

Fig. 43.   Fig. 44.

Fig. 43. Development of SER features from C_2H_2 on coldly evaporated silver film with exposure. Note change of intensity scale. 200 mW of 514.5 nm radiation, 4.5 cm^{-1} bandpass. After /525/

Fig. 44. Annealing of acetylene SER features on coldly evaporated silver film. (a): Spectrum from exposed surface (36 L), T = 120 K (top); same, but annealed to 170 K and recooled to 120 K (bottom); (b): detail from spectrum at bottom in (a); (c): low energy features after annealing to 170 K (C_2H_2 exposure!). 200 mW of 514.5 nm radiation, 4.5 cm^{-1} bandpass. After /525/

(0.036 L; Fig. 43). As expected, the line at the high energy side develops to the most prominent feature for exposures approaching saturation.

Annealing Behaviour. If an exposed sample is annealed to 170 K and subsequently re-cooled to 120 K, the spectrum at the bottom of Fig. 44a is observed (C_2D_2 exposure). The lines attributed to adsorbed C_2D_2 have disappeared due to desorption of this species at $\approx$ 145 K (desorption of C_2H_2 from Ag(110) has been observed between 100 K and 160 K /522/), bands due to adsorbed C_2D are still present. The band assigned to ν_{C-C} of C_2D consists of several lines (Fig. 44b), presumably caused by bonding to different adsorption sites (see discussion in Sect. 5.1.2). The low frequency region is dominated by a broad peak centered at $\approx$ 270 cm^{-1} (Fig. 44c; C_2H_2 exposure), which is attributed to the metal-acetylide stretching mode. Note also, that the band at $\approx$ 2110 cm^{-1} has gained intensity. Upon further annealing, monodentate acetylide bands disappear between $\approx$ 170 K and $\approx$ 220 K. Simultaneously, further intensity increase of the band at $\approx$ 2110 cm^{-1} is observed. This line as well as two bands at $\approx$ 230 cm^{-1} and $\approx$ 330 cm^{-1} are the only features still seen in the SER spectra at $\approx$ 230 K. Following /522/ we assume, that C_2H (C_2D) has disproportionated to adsorbed bidentate acetylide C_2: remaining SER lines may be attributed to ν_{C-C} and ν_{Ag-C_2} (symmetrical and antisymmetrical). We note again, that these bands are frequently observed in SER spectra from coldly evaporated Ag films, e.g. after extended CO exposure. A detailed discussion of these features is therefore postponed to the following chapter.

General Discussion. Acetylene adsorption has so far been studied by surface enhanced Raman spectroscopy only on silver surfaces. Breakdown of ordinary Raman selection rules as well as mode selective enhancement has been observed. This may be interpreted following the scheme outlined for ethylene (Sect. 5.1.2). We emphasize again, that, similar to metal-olefin complexes /479/, drastic changes of the intensity distribution in Raman spectra of some metal organo-acetylides compared to their parent acetylides have been observed /527/ (for coordinated alkynes (alkenes), bands due to triple (or double) C-C bond stretching are much more pronounced relative to those of other modes, e.g. of C-H stretching vibrations /479,489,527/. As acetylene bonding in complexes /527/ and to silver surfaces /522/ displays some similarities, SER spectra from C_2H_2 on Ag are expected to be similarly affected. Any attempt to understand mode selective enhancement in SER spectra should consider these effects. SER line intensities should be compared to those of appropriate complexes rather than to those of the isolated molecule, when elaborating on the contribution of the SER process itself to the mode selective enhancement.

Three species can be distinguished in SER spectra from C_2H_2 on Ag. Weakly, probably π bonded C_2H_2 ($T_{des} \approx$ 145 K), σ bonded monodentate acetylide C_2H (additional *weak* π- or μ bonding is possible /522,528/; $T_{dispro} \approx$ 170 - 220 K), and di-σ bonded bidentate acetylide C_2 /522/. SER data suggest, that the C-C triple bond is only negligibly perturbed in all three species. Vibrational bands consist of several lines (ν_{C-C}) or are inhomogeneously broadened, which is not surprising regarding the inhomogeneity of the surface of coldly evaporated silver films. Adsorption on sites with different local environment can account for these observations (see also Sect. 5.1.2). Surfaces exposed and investigated at 120 K display SER features of all three species, where those from C_2H_2 and C_2H are of comparable intensity (C_2 features are much weaker). Upon annealing, C_2 bands develop at the expense of C_2H bands until the latter have disappeared (at $\approx$ 220 K). As mentioned, similar effects have been observed on Ag(110) /520-522/. At $\approx$ 100 K, adsorbed acetylene reacts with atomic oxygen to form C_2H, which disproportionates into adsorbed C_2 at $\approx$ 300 K. Our observations may be explained correspondingly. Recalling the fondness of Ag defect surfaces for oxygen adsorption (see Chapt. 7), it is not entirely inconceivable that some oxygen atoms are present on our vapour deposited silver films (basic properties of the silver/oxygen system are outlined in Chapt. 7). Alternatively, partial dehydrogenation of C_2H_2 to C_2H may be rendered possible by "special defect sites" not present on smooth Ag(110) (dehydrogenation of C_2H_2 on stepped Ni surfaces is well established /436,515/). We favour the latter interpretation, but, clearly, available data do not allow any final conclusion. Further vibrational studies could help to clear the situation. Particular meaningful experiments are SER studies on silver films pre-coated with atomic oxygen or EELS studies on single crystal faces of Ag with known defect structure (steps, kinks, etc.).

5.2 Cyclic Hydrocarbons

5.2.1 Benzene

The benzene molecule is fairly well characterized. It has a planar structure with the carbons forming a regular hexagon (symmetry group D_{6h}). The molecule has twenty fundamentals (ten are doubly degenerate), of which fourteen are in-plane modes ($2A_{1g} + 1A_{2g} + 2B_{1u} + 2B_{2u} + 3E_{1u} + 4E_{2g}$) and six are out-of-plane modes ($1A_{2u} + 2B_{2g} + 1E_{1g} + 2E_{2u}$). Seven modes are Raman active ($2A_{1g} + 4E_{2g} + 1E_{1g}$), four are infrared active ($1A_{2u} + 3E_{1u}$), and the remaining nine are inactive. Vibrational data may be found in /1, 467, 529,530/ (in this review the notation of /531/ for the benzene fundamentals is adopted). Since the molecule does not absorb in the visible, only ordinary Raman scattering is expected. Some modes have very large scattering cross sections /19/, especially in benzene derivatives like $C_6H_5NO_2$. Optical absorption starts with a band at 4.65 eV due to a forbidden electronic transition made possible by vibronic interaction /424/ (the first allowed transition leads to an intense band at 6.96 eV). Electron impact spectra reveal an additional peak at 3.95 eV due to the doubly forbidden lowest singlet-triplet transition.

Various benzene organo-metallic complexes have been synthesized /474,475/, of which dibenzenechromium $Cr(C_6H_6)_2$ is particularly well characterized (vibrational spectra are discussed in detail in /532 - 535/). This complex has a sandwich structure, with the metal bonded to the π-electron system of the two benzene rings. The compound adsorbs in the visible /536/. Complexes of C_6H_6 with the noble metals of group Ib are also known /475/, e.g. $AgClO_4 \cdot C_6H_6$ /537/, but are less comprehensively investigated /480/.

A variety of surface sensitive techniques such as LEED, UPS, TDS, AES, and $\Delta\phi$ measurements have been employed to study benzene adsorption on metals (see, e.g., /325, 357,367,454,506,538 - 540/). Vibrational properties have been probed by neutron inelastic scattering (C_6H_6 on Raney nickel /541/), by fourier transform infrared spectroscopy (C_6H_6 on alumina-supported Pt /542,543/), as well as by high resolution EELS (C_6H_6 on Ni(100) /544,545/, on Ni(111) /544 - 546/, and on Pt(111) /546/). Vibrational data of C_6H_6 on Ag(111) have also recently been reported /547/. It is concluded, that benzene adsorption proceeds non-dissociatively on the investigated surfaces (at room temperature). Bonding through the π electrons of the ring with the aromatic ring parallel to the surface is very likely. As inferred from the remarkable similarity of vibrational frequencies of adsorbed and liquid benzene, interaction of C_6H_6 with Ag(111) is comparatively weak /547/. For the sake of completeness we mention, that σ bonding of C_6H_6 has been observed on a Pt/Al_2O_3 catalyst surface in the presence of a structured carbon residue /543/.

Several surface enhanced Raman vibrational studies of benzene adsorption on silver have been published /50,116,118,133/. SER line frequencies from C_6H_6 on coldly evaporated Ag films (T = 11 K) differ only slightly from those of the isolated molecule (Table 10). The out-of-plane bending ν_{11} (+ 24 cm^{-1}) and the symmetric breathing vibration ν_1 (-10 cm^{-1}) display the most noticable shifts. Interestingly, qualitatively similar behaviour is found for π-complexed benzene and benzene adsorbed on single crystal surfaces (Table 10; note particularly the excellent agreement of SER and $AgClO_4 \cdot C_6H_6$ vibrational data). This suggests similar benzene-metal interaction in all these systems. Hence, it is concluded, that C_6H_6 lies flatly on the silver film forming a weak π bond. Raman spectra from silver films deposited at 300 K on an optical grating and subsequently exposed at 100 K to C_6H_6 display a weak ν_1 band at 984 cm^{-1} only for coverages far below a monolayer /50/ (scattering under plasmon surface polariton resonance conditions). Its intensity saturates at a fraction of a monolayer, suggesting that it is associated with adsorption on some particular site. For higher coverages ($\gtrsim$ monolayer) a line at 991 cm^{-1}, close to the gas phase value of ν_1, dominates the spectrum. From benzene on a sinusoidally modulated Ag(111) surface (amplitude $\approx$ 50 nm, period $\approx$ 1000 nm) no Raman signal has been obtained until $\approx$ 8 monolayer equivalents were on the surface /547/. The multilayer Raman spectrum displayed ν_1 at 990 cm^{-1}. These results seem to indicate, that strong SERS is associated with benzene on some particular adsorption site on coldly evaporated films,

Table 10. Vibrational frequencies of benzene in different environment. (a): gas (liquid) phase data of C_6H_6, after /467/; (b): dibenzenechromium, after /532,534/ (the first value is for the in-phase, the second for the out-of-phase mode of the two benzene rings); (c): $AgClO_4 \cdot C_6H_6$ in solution, after /480/; (d): C_6H_6 on Ag(111), T = 100 K, after /547/; (e): C_6H_6 on Pt(111), T = 140 – 320 K, after /546/; (f): SERS from C_6H_6 on coldly evaporated silver film, T = 11 K, after /133/

Mode Symmetry	(a) Gas (Liquid) C_6H_6	(b) Coordinated $Cr(C_6H_6)_2$	(c) $AgClO_4 \cdot C_6H_6$	(d) EELS Ag(111) C_6H_6	(e) EELS Pt(111) C_6H_6	(f) SERS Ag Film C_6H_6
$\nu_{Me-C_6H_6}$		459			360/570	
ν_{16}, E_{2u}	405	409/400				397
ν_6, E_{2g}	607	604/630	602			605
ν_{11}, A_{2u}	673	794/791		675	830/920	697
ν_4, B_{2g}	690					
ν_{10}, E_{1g}	850	811/833				864
ν_{17}, E_{2u}	970	910/910				970
ν_5, B_{2g}	984	950				
ν_1, A_{1g}	992	970/971	980	1000		982
ν_{12}, B_{1u}	1006					
ν_{18}, E_{1u}	1038	999/999				1032
ν_{15}, B_{2u}	1146	1142		1155	1130	1149
ν_9, E_{2g}	1178	1143	1175			1174
ν_{14}, B_{2u}	1310	1308			1420	1311
ν_3, A_{2g}	1326					
ν_{19}, E_{1u}	1486	1426/1430		1480		1473
ν_8, E_{2g}	1586	1631/1592	1589			1587
ν_7, E_{2g}	3047	2955				
ν_{13}, B_{1u}	3057	2855				
ν_2, A_{1g}	3062	3053/3053	3065	3030	3000	3060
ν_{20}, E_{1u}	3080	2904/2904				

whose density is comparatively small on room temperature deposited films and which is absent on Ag(111) (compare to pyridine SER features, Chapt. 4).

Ordinary Raman selection rules are relaxed, and mode selective enhancement is observed in SER spectra of benzene /116,133/. Reduction of the molecule's symmetry from D_{6h} to $C_{3v}(\sigma_d)$ (or lower) upon adsorption would account for the former /116/. This corresponds to a benzene molecule lying flat atop an equilateral triangle of silver atoms with the normals to three alternate edges of C_6H_6 pointing to the centers of the three atoms, which is a quite reasonable adsorption geometry on a (111) surface /546,547/ (evaporated silver films expose mainly (111) faces, see Sect. 4.4). Alternatively, it has been suggested /116,133/, that a large field gradient at the metal surface would change the selection rules of C_6H_6 such, as if the symmetry was reduced from D_{6h} to $C_{3v}(\sigma_d)$, independent of the local geometry. Yet another possible explanation based on a particular charge transfer model for the chemical enhancement has been discussed in /67/. As with other SER spectra (see preceeding sections), any quantitative valuation of the mode selective enhancement should include bonding induced changes of spectral features.

SERS from benzene on coldly evaporated lithium /133/ and sodium films /142/ has also been reported (T = 11 K and 15 K, respectively). SER vibrational energies are similar to those from C_6H_6 on silver. A detailed analysis of the spectra has, however, not been performed. It is remarkable, that the relative line intensities in SER spectra from Li are quite different from those of Ag /133/. Published SER spectra from C_6H_6 on mercury /149/ could not be reproduced by other groups (see Chapt. 2), and it is likely, that the conlcusions drawn in /149/ have to be corrected. Benzene on silica-supported Ni /78/ and Pt as well as on Pt clusters /156/ (diameter $\approx$ 10 nm) displayed Raman spectral features similar to those from corresponding single crystal faces /544 - 546/. In addition, many usually Raman forbidden lines were observed, which has been explained with lowering of the molecular symmetry by the site symmetry of the surface as sketched above. For details of these very interesting results the reader is referred to the original papers /78,156/. To the opinion of the author, the estimated enhancement factors ($10^3 - 10^4$) are debatable (see also Chapt. 2).

5.2.2 Benzene Derivatives

Benzene derivatives (carboxylic acids) have been used in basic SER studies for mainly two reasons. Firstly, some vibrational modes of these molecules have a very large ordinary Raman scattering cross section /19/, and, secondly, from solution they are deposited in a known geometry on oxide (or sulfide) surfaces /62/ (via chemisorption of the carboxylate group; note, that all experiments described below have been performed with the sample in air, where an oxide or sulfide layer is present on silver).

Benzoic Acid. Strongly enhanced Raman spectra from benzoic acid on silver island films prepared on glass substrates have only been observed, when the molecule was chemisorbed to the metal via the carboxylate group /62,294/. When the molecule was first deposited on the glass substrate and subsequently overcoated with the island film, so that the benzene ring was in close proximity to the silver, no Raman signal was observed. This has been taken as evidence for the importance of chemisorption in the enhancement process /294/. Vibrational analysis of the data has not been performed.

Nitrobenzoic Acid. Intense, enhanced Raman spectra from a monolayer of nitrobenzoic acid spin-deposited on silver island films have been observed /44,548,549/. An enhancement factor of $\approx 10^5$ has been estimated for the ring mode at 1596 cm^{-1} /548/. The absence of any SER feature in the 1700 cm^{-1} region corresponding to vibration of the COO-H group provided evidence for chemisorption via the carboxylate group /44/. Again, analysis of the SER vibrational features has not been presented. As demonstrated recently, a sub-monolayer of nitrobenzoic acid on *smooth* aluminum surfaces yields detectable, *unenhanced* Raman signals when using multichannel optical detectors /550,551/.

Aminobenzoic Acid. The detailed analysis of SER spectra from a monolayer of various n-aminobenzoic acids (n = 2,3,4) on silver island films has been postponed to a later publication /552/. It is interesting, that 4-aminobenzoic acid displayed a highly structured SER spectrum, whereas only some broad bands were observed from 2-aminobenzoic acid.

Trifluorobenzene. SER spectra from 1,3,5-trifluorobenzene on coldly evaporated silver films (11 K) change drastically after several minutes warm up at 150 K /133/. Lines of the halogenated benzene are considerably reduced in intensity and several new lines grow in. An analysis of the spectral changes to identify the products has not been performed. The observations demonstrate, however, the possible utility of SERS in following surface reactions.

Finally, we note the report of SERS from cyclohexane C_6H_{12} on iodine-roughened silver surfaces in UHV /553/. Unfortunately, no experimental details have been published.

5.3 Summary

Adsorption of various hydrocarbons on several high reflectivity metals (mainly silver) has been studied by SERS. No surface enhanced spectra from coldly evaporated silver films could be observed for the two alkanes investigated, namely methane and ethane. On the other hand, signals from adsorbed hydrocarbons with unsaturated C-C bonds are appreciably enhanced (ethylene, propylene, butene, acetylene, benzene).

Adsorption on silver of the two unsaturated C_2 hydrocarbons has most comprehensively been investigated.

SER signals from ethylene on coldly evaporated silver films are very strong (ν_3, ν_2): only 0.1 per cent of a monolayer is easily detected. As indicated by the SER spectral features, the molecule is only weakly adsorbed by π interaction (presumably to positively charged sites only). Desorption of this species is observed between 160 K and 185 K. New SER lines appear in this temperature range. These are attributed to an adsorbed bidentate acetylide species, probably from decomposition of part of the adsorbed ethylene during warming up.

Three species are identified in SER spectra from silver exposed to acetylene. At 120 K, a weakly π-bonded species ($T_{des} \approx 145$ K) coexists with adsorbed monodentate acetylide C_2H. This points to the existence of special sites on coldly evaporated silver films which are able to partly dehydrogenize acetylene upon adsorption at 120 K. Lines attributed to C_2H disappear between $\approx$ 170 K and $\approx$ 220 K. Simultaneously, new bands assigned to adsorbed bidentate acetylide grow in: C_2H is apparently further dehydrogenized to C_2 in this temperature range. The nature of the "special active" adsorption sites, which allow the dehydrogenation of acetylene, is not clear at present. Adsorbed impurities (oxygen atoms) or special defect sites might be important. We note, that the bands assigned to the C_2 species are frequently observed in SER spectra from coldly evaporated silver films (e.g. after annealing of C_2H_4 exposed samples or after extensive CO exposure, see following chapter).

The hydrocarbon adsorption studies convincingly demonstrate the detailed spectral information obtainable with SERS and its very high sensitivity for certain adsorbates on silver. The utility of this technique to follow surface reactions of adsorbed species is evident from the presented results.

6. Carbon Monoxide Exposure and Carbonaceous Deposits

The great, constant interest in carbon monoxide adsorption on metal surfaces is caused by essentially two reasons. Firstly, the relative simple interaction of CO with a metal can be regarded as a model system for basic studies of chemisorption. Secondly, there are several metal-catalyzed CO reactions of industrial importance such as the catalytic methanation /554 - 556/, the Fischer-Tropsch synthesis /557, 558/, or the remove of CO from automobile exhaust gases /559/. As a consequence, a vast amount of literature on CO adsorption on metals has been published. Several review articles or chapters in textbooks give an introduction into the field (see, e.g., /13,16,560,561/).

The structure and properties of the isolated molecule are well known /562/. The carbon-oxygen distance is 1.1282 $\mathring{A}$, and the molecule has a small permanent dipole moment of 0.112 D /563/. It does not absorb light in the visible, and the stretching frequency ν_{C-O} appears as a very strong band due to the large dynamical dipole moment at 2143 cm^{-1} in infrared spectra. Carbon monoxide forms complexes with almost every transition element as well as with metals of group Ib. Spectroscopic data are summarized in several books (e.g. /474,564/). Spectral properties of copper and silver carbonyls are discussed in some detail in /565 - 568/. Due to metal-to-ligand charge transfer excitations, many metal carbonyls absorb in the near ultraviolet or visible frequency range /564/.

CO adsorption on metal surfaces has extensively been studied with various surface sensitive techniques. Vibrational data are compiled in /13,16,560,569,570/. Interaction with the noble metals of group Ib, especially with silver, is very weak. Angle resolved photoemission data point to only physisorption on Ag(110) /571/ (adsorption on Au is treated in, e.g.,/572 - 576/, interaction with Ag in /572,574,575, 577 - 582/; the more numerous studies on Cu surfaces are summarized in a recent review article /583/, where further references can be found).

The generally accepted picture of carbon monoxide bonding to metals is by electron transfer from the 5σ orbital of CO to the metal and by backdonation of metallic electrons into the empty, antibonding $2\pi^*$ orbital of the molecule /564/. This weakens the bond between carbon and oxygen and hence leads to a decrease of ν_{C-O} below

the gas phase value. The magnitude of the downward shift depends on the details of
the bonding to the metal. Therefore the value of the C-O stretching frequency can be
used to identify metal-CO bonding properties, and the following assignment scheme
based on results of /584,585/ is generally accepted /560/: (i) ca. 2200 cm^{-1} to
ca. 2130 cm^{-1}: CO on oxidized metals, e.g. Cu^{2+}; (ii) ca. 2130 cm^{-1} to ca. 2000 cm^{-1}:
linearly bonded CO; (iii) ca. 2000 cm^{-1} to ca. 1880 cm^{-1}: CO bridging *two* metal
atoms; (iv) ca. 1880 cm^{-1} to ca. 1650 cm^{-1}: CO bridging *three or more* metal atoms.

Carbonaceous deposits are also treated in this chapter, since SER spectral features from CO exposed silver surfaces are often obscured by various "impurity bands"
which are mainly caused by carbon containing species (see, e.g.,./61/).

6.1 Adsorbed Carbon Monoxide

Raman signals have been obtained from CO dosed coldly evaporated noble metal films
/61,134,136,281,360,370,586 - 588/, colloidal silver particles isolated in a solid
CO matrix /132/, and from low index faces of Ni and silica-supported Ni exposed to
CO /79,82,159,161/. Spectra from extensively exposed silver films (up to 10^5 L,
T = 120 K, /61,134/) display many lines between 40 cm^{-1} and 2600 cm^{-1}. These are
certainly enhanced Raman features, but they are most likely due to some impurity
adsorbates rather than to adsorbed CO. Carbon monoxide does not remain on silver surfaces at liquid nitrogen temperature when the chamber is evacuated /370/. Likewise,
lines at $\approx$ 2110 cm^{-1} in spectra from CO on Ag colloids /132/, from thick layers condensed on Ag films (T = 11 K, /136/), or from extensively dosed Ag films (T = 120 K,
/587/) are presumably not due to adsorbed carbon monoxide (see below).

Coldly evaporated silver films (T = 120 K) in contact with an ambient CO pressure
display a *single* strong Raman line at 2135 cm^{-1} close to the gas phase value of ν_{C-O}
/586/ (Fig. 45; the low energy band at 160 cm^{-1} is assigned to ν_{Ag-CO} in /586/). The
Raman enhancement was estimated to $\approx$ 10^4 - 10^5 /586/. When the ambient pressure is
raised, i.e. CO coverage increased, ν_{C-O} shifts to smaller energy [Fig. 45, inset;
for the smallest pressure, the C-O stretching frequency is practically identical to
the gas phase value; note, that, even for 10^{-3} Torr, CO coverage is far below saturation (at 120 K, Ag(111), /580/)]. This shift of ν_{C-O} with coverage is frequently observed (usually upward /589 - 592/, sometimes downward /574,575,593/). Several mechanisms might be involved /464,594 - 599/. Because of the very weak interaction of CO
with silver, it seems reasonable to neglect substrate induced chemical effects. Dipole-dipole coupling between the vibrating molecules is likely to be most important.
Usually, this interaction results in an upshift of ν_{C-O}, since the molecules vibrate
perpendicular to the surface for common adsorption geometries of CO. The observed
downshift might be taken as support for the adsorption geometry proposed in /571/,
namely an orientation of the molecular axis parallel or slightly inclined to the

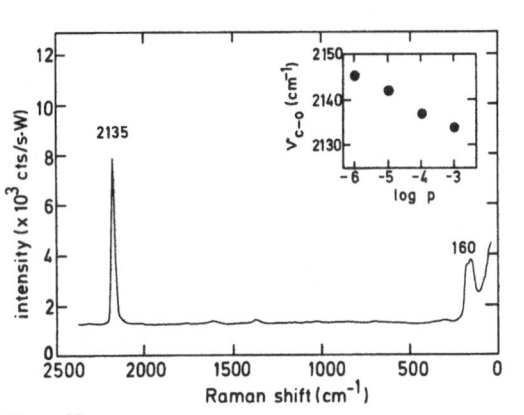

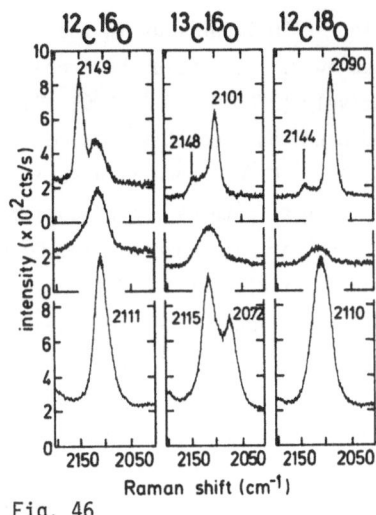

Fig. 45

Fig. 46

Fig. 45. SER spectrum from coldly evaporated silver film (T = 120 K) in contact with $5 \cdot 10^{-4}$ Torr carbon monoxide. 300 mW of 488.0 nm radiation, 4 cm^{-1} bandpass. Inset shows downshift of ν_{C-O} with increasing pressure. After /586/

Fig. 46. SER spectra from coldly evaporated silver films (T = 120 K) exposed to various carbon monoxide isotopes as indicated. Top rank: sample in contact with $1 \cdot 10^{-6}$ Torr carbon monoxide; middle rank: after measurement of top rank spectra, chamber evacuated [integrated dose was 800 L ($^{12}C^{16}O$) and 700 L ($^{13}C^{16}O$ and $^{12}C^{18}O$)]: bottom rank: sample exposed to $3 \cdot 10^{4}$ L carbon monoxide, chamber subsequently evacuated. All spectra have been taken with 200 mW of 514.5 nm radiation and 4 cm^{-1} bandpass. After /587/

surface. Dipole coupling between molecules vibrating parallel or almost parallel to the surface shifts ν_{C-O} to smaller values (analogous to the redshift of the transverse collective electron resonance with respect to the single particle resonance /196/ described in Chapt. 2). It is clear, that an exactly parallel adsorption geometry would be incompatible with the fact, that ν_{C-O} is observed in infrared spectra from CO an Ag(100) /581/. The given, tentative interpretation is certainly not more than a crude approach to the problem. It neglects, for instance, any possible contribution of coadsorbed impurity molecules to spectral shifts /595,600/. Further studies are necessary before final conclusions can be drawn.

The assignment of the SER line at 2135 cm^{-1} to ν_{C-O} /586/ is confirmed by isotope experiments /587/ (Fig. 46, top rank, $1 \cdot 10^{-6}$ Torr CO). The observed displacements of ν_{C-O} with respect to the frequency of $^{12}C^{16}O$ compare very well to calculated shifts: $^{13}C^{16}O$: 48 cm^{-1}(obs.) / 49 cm^{-1}(calc.); $^{12}C^{18}O$: 59 cm^{-1}(obs.) / 53 cm^{-1}(calc.). Bands at $\approx$ 2145 cm^{-1} in the spectra from the carbon-13 and oxygen-18 labeled compounds are due to $^{12}C^{16}O$ impurities in the gas. After the measurements, the carbon monoxide was pumped away. The samples had then accumulated doses of 800 L ($^{12}C^{16}O$) and 700 L respectively ($^{13}C^{16}O$ and $^{12}C^{18}O$). Raman spectra from these silver surfaces display a broad band centered at $\approx$ 2110 cm^{-1}, *independent of the isotope used* (Fig. 46, middle rank; for $^{13}C^{16}O$, there is a second weak line on the low energy side of the main band). These features are discussed in Sect. 6.2. In agreement with the results presented so far, SER spectra from CO condensed on Ag films at 10 K under UHV conditions show ν_{C-O} at 2142 cm^{-1} /370/.

SER vibrational data are compared to infrared results in Table 11. Only bands between 2100 cm^{-1} and 2200 cm^{-1} are considered. For CO on evaporated films and Ag(100), ν_{C-O} is quite consistently found very close to the gas phase value. This documents the negligible perturbation of the molecule upon adsorption. It seems reasonable to assume that the molecule is only physisorbed as suggested in /571/. CO on oxidized silver surfaces as well as silvercarbonyl ions in solution display ν_{C-O} at considerably higher frequency ($\approx$ 2180 cm^{-1}, Table 11).

Table 11. CO stretching frequencies for carbon monoxide in various environments

ν_{C-O} (cm^{-1}) 2200 — 2150 — 2100	Technique	Sample	Reference
(marker near 2140)	SERS	10^{-6} - 10^{-3} Torr, film, 110 K	/568/
(marker near 2150)		condensed, film, 10 K	/370/
(marker near 2155)		10^{-6} Torr, film, 120 K	/587/
(marker near 2145)	INFRARED	$7.5 \cdot 10^{-4}$ - $7.5 \cdot 10^{-2}$ Torr, film, 113 K	/574/
(marker near 2135)		$5 \cdot 10^{-6}$ - 1.5 Torr, film, 77 K	/575/
(marker near 2145)		10^{-8} - 10^{-3} Torr, Ag(100), 80 K	/581/
(markers near 2155, 2140)		0.3 Torr, reduced Ag/SiO$_2$, 160 K	/578/
(marker near 2170)		1 Torr, Ag/Al$_2$O$_3$, 160 K	/572/
(marker near 2175)		20 Torr, oxidized Ag/SiO$_2$, 300 K	/577/
(marker near 2180)		20 Torr, oxidized Ag/SiO$_2$, 300 K	/578/
(markers near 2185, 2175)		Ag(CO$_2$)$^+$	/566,568/
(marker near 2155)		isolated CO	/562/

Compared to silver, the interaction of CO with gold and copper is considerably stronger /571,574/. Raman spectra from coldly evaporated gold films, which had been exposed to 10^5 L CO, display a single, weak line at 2118 cm^{-1} /134,586/ (T = 120 K, green light excitation!). As the frequency agrees with infrared data from CO on gold films (ν_{C-O} = 2115 cm^{-1} - 2120 cm^{-1}, /560/), the line is attributed to the C-O stretching vibration of adsorbed CO. Further investigations are necessary to consolidate the assignment, red light excitation should give better performance (see Chapt. 2). Spectra from coldly evaporated Cu films exposed to 30 L of CO show three lines at 2104 cm^{-1}, 355 cm^{-1}, and 285 cm^{-1} /588/ (Fig. 47; red light excitation, T = 120 K). Practically identical spectra were obtained for smaller exposures. Saturation of the Raman features was observed for a dose of several Langmuir /588/. At the coverage certainly does not exceed a monolayer, the spectra are enhanced. The lines are tentatively assigned to ν_{C-O}, ν_{Ag-CO}, and the bending mode δ_{C-O} (frustrated rotation) of linearly bonded CO. Support for this interpretation comes from the following arguments. (i) The frequency of ν_{C-O} agrees very well with that found in numerous infrared studies of CO on evaporated Cu films (ν_{C-O} = 2102 cm^{-1} - 2107 cm^{-1}, /560,583/). (ii) Less data are available for the other modes. Consistent with the SER value, ν_{Cu-CO} has been observed at 343 cm^{-1} (EELS study, Cu(100), T = 80 K, /601,602/) and in the range ≈ 325 cm^{-1} - 375 cm^{-1} (matrix-isolated coppercarbonyl, /565/). Experimental data for δ_{C-O} have not been found. Following the path outlined for CO on nickel in /603/, a frequency of δ_{C-O} close to the observed value is estimated. (iii) At 80 K, CO saturation coverage is obtained after an exposure of several Langmuir /601, 604/. A similar dose is sufficient to saturate SER features /588/. (iv) Desorption of CO from polycrystalline Cu surfaces is completed at ≈ 200 K. SER features are lost at ≈ 210 K /588/. SER studies using different CO isotopes could further prove the given interpretation. It is at present unclear, why the low energy modes are so strong (or, vice versa, why the stretching mode is so weak) in SER spectra from CO exposed Cu films.

Raman spectra from carbon monoxide on Raney nickel /82/, silica-supported Ni /82, 159,161/, and some low index faces of Ni /79/ display a variety of lines in the region

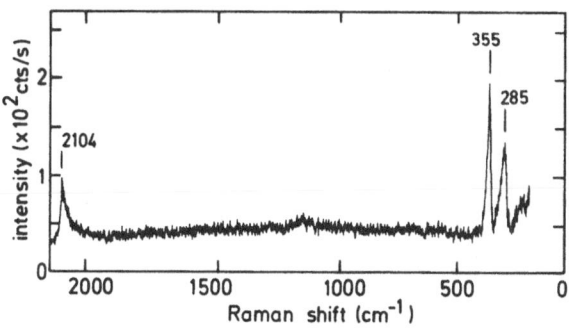

Fig. 47. SER spectrum from coldly evaporated copper film (T = 120 K) exposed to 30 L of carbon monoxide. 200 mW of 647.1 nm radiation, 4.5 cm^{-1} bandpass. After /588/

of the C-O stretching vibration as well as at lower frequencies. Some of the spectra are claimed to be surface enhanced /161/. As outlined earlier (Chapt. 2), this conclusion needs further proof. For spectral details the reader is referred to the original literature.

6.2 Carbonaceous "Impurity" Deposits

SER spectra from coldly evaporated silver films in contact with $1 \cdot 10^{-6}$ Torr carbon monoxide display a broad band at the low energy side of ν_{C-O} (Fig. 46; middle rank). After extensive CO exposure, it is detected as a strong band centered at ≈ 2110 cm^{-1}, *independent of the isotope* used for the exposure (Fig. 46; bottom rank). The carbon-13 labeled compound shows a second less intense line at 2072 cm^{-1}, which is accompanied by a third very weak and hardly visible line at ≈ 2030 cm^{-1}. The lines are considerably broader than those usually observed in SER spectra ($\Delta\nu \approx 40$ cm^{-1}, compare to, e.g., Fig. 12). The development of these features with exposure is shown in Fig. 48. The CO pressure in the vacuum chamber was set to $1 \cdot 10^{-6}$ Torr, and spectra were taken after different time intervals. With increasing time, i.e. increasing total, integral exposure, the band at 2110 cm^{-1} grows at the expense of the ν_{C-O} line. This indicates "poisoning" of the silver surface by the species responsible for the line at 2110 cm^{-1}. It is assumed, that it covers an increasing part of the surface with increasing exposure thus reducing the area available for CO adsorption. Consequently, the SER signal from adsorbed carbon monoxide decreases.

In addition to the features around 2100 cm^{-1} (Fig. 46), a variety of lines are observed after extensive CO exposure (Fig. 49; chamber evacuated). As these often accompany SER spectra from adsorbed molecules on silver as so-called "impurity" lines (see, e.g., Chapt. 5), an identification of these features seems desirable. A tentative interpretation scheme based on the following arguments is given in Fig. 49. Practically identical spectra were recorded for all three isotopes. Only the spectrum taken after $^{13}C^{16}O$ exposure displays a few additional lines between 1850 cm^{-1} and 2100 cm^{-1}. This argues against an interpretation based on adsorbed, *oxygen containing*, CO-derived species supplied with the gas feed or reactively synthesized at the silver surface or any other part of the vacuum chamber (e.g. against carbonyls or carbonate). It rather points to an impurity, which is almost equally offered with every isotope. Comparison of the spectral features to SER spectra of ethylene and acetylene (Chapt. 5) suggests, that unsaturated hydrocarbons play a leading role. The dominating band at 2113 cm^{-1} is assigned to the stretching mode of the completely dehydrogenized bidentate acetylide C_2, with $^{13}C^{12}C$ and $^{13}C_2$ being responsible for the lines at 2072 cm^{-1} and ≈ 2030 cm^{-1} respectively. As estimated from the different masses, the latter two are expected at 2072 cm^{-1} and 2030 cm^{-1} in good agreement with the observation. The silver-acetylide stretching modes (symmetric and asymmetric)

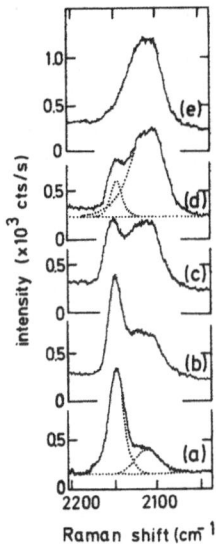

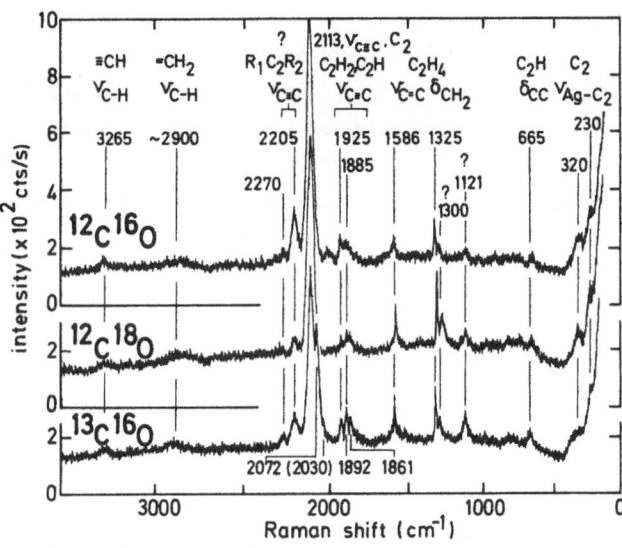

Fig. 48

Fig. 49

Fig. 48. Development with time of SER features from coldly evaporated silver films (T = 120 K) in contact with 1 10^{-6} Torr carbon monoxide. (a): after 6.5 min (390 L); (b): after 20 min (1200 L); (c): after 33 min (1980 L); (d): after 46 min (2760 L); (e): after 48 min (2880 L), chamber evacuated. 200 mW of 514.5 nm radiation, 4 cm^{-1} bandpass. After /587/

Fig. 49. SER spectra from impurities on coldly evaporated silver films (T = 120 K) after exposure to $3 \cdot 10^{4}$ L of various carbon monoxide isotopes. Tentative assignment of spectral features is indicated in the top part, energies of additional features after $^{13}C^{16}O$ exposure are given at the bottom. 200 mW of 514.5 nm radiation, 4 cm^{-1} bandpass. After /587/

give rise to the features below 350 cm^{-1}. Lines between 1850 cm^{-1} and 1940 cm^{-1} are attributed to adsorbed acetylene and monodentate acetylide C_2H, with additional features from carbon-13 labeled compounds in the bottom spectrum (Fig. 49). Further support for this interpretation is provided by the annealing behaviour of these lines /605/, which parallels that observed for acetylene (see Chapt. 5). Frequencies of two sharp lines (1586 cm^{-1} and 1235 cm^{-1}) are in excellent agreement with those of the C-C stretching and symmetric scissors mode of adsorbed ethylene (see Table 4). C-H stretching modes of this species contribute to the weak broad feature at $\approx$ 2900 cm^{-1} (corresponding modes of acetylenic species give the band at 3265 cm^{-1}). The origin of several further bands between 1950 cm^{-1} and 2300 cm^{-1} is uncertain. They might be due to C-C stretching modes of substituted acetylenic (> 2100 cm^{-1}) and allenic species (< 2000 cm^{-1}) /6/.

In closing we again point at the preliminary character of the given interpretation. The assignments are supported by the good agreement of vibrational frequencies with those of identified adsorbates, and by the fact, that the species found at the surface are usually present in the background pressure of ion getter pumped vacuum systems. Their partial pressure might be considerably higher during CO exposure due to corresponding impurities in the supply. SER spectra from extensively dosed cold silver surfaces or from cold samples exposed in relatively poor ambient background pressure (e.g. /133/) have therefore carefully to be checked of impurity lines from hydrocarbon derived deposits. Even trace amounts of these impurities may lead to appreciable SER signals (see, e.g., Fig. 46).

6.3 Amorphous Carbon

Many Raman spectra from nominally clean silver surfaces display a structured, broad and quite intense band between 1000 cm^{-1} and 1600 cm^{-1} (e.g. /25,96,266/; the band was first described in /90/). Fig. 50a shows a spectrum from a polished polycrystalline silver slug /25/, which was cleaned by several sputtering-heating cycles under UHV conditions until the surface was free of contaminants as checked by AES (free within the sensitivity of this technique; Fig. 50b displays the corresponding Auger spectrum). Since the molecule responsible for the Raman bands escapes detection with Auger, it is clear, that the measured signal is unusually strong (a contribution to the band of subsurface molecules within the penetration depth of the light is also conceivable as outlined in /25/). Similar Raman features have been obtained from island films /266/, chemically roughened Ag surfaces /366/, and Ag(111) faces with inscribed grating /48/ (all prepared in UHV). They are absent in spectra from appropriately prepared, coldly evaporated silver films. However, after warming the latter to room temperature, they slowly develop, even under UHV conditions /606/. Note, that these Raman features are stable at room temperature.

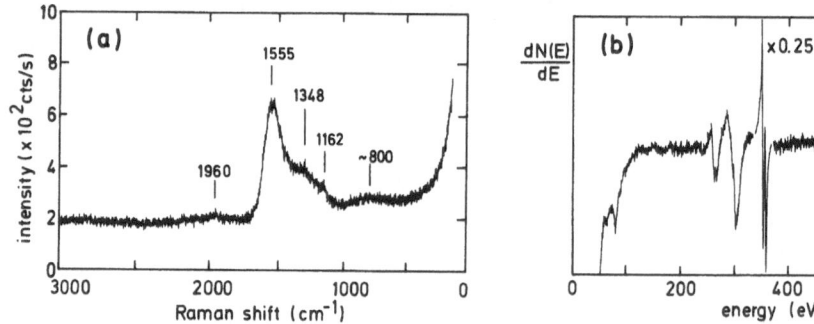

Fig. 50. (a): Raman spectrum from polycrystalline, sputter-cleaned silver surface (T = 300 K; 250 mW of 514.5 nm radiation, 3 cm^{-1} bandpass). (b): Auger spectrum of the same surface (E_0 = 2 keV, mod.: 3 V_{pp}). After /25/

A detailed Raman and EELS study of various carbonaceous deposits on silver /607/ revealed, that the bands described above are due to graphitic carbon contaminants below the detection limit of AES. The Raman spectrum of crystalline graphite is dominated by a single band at 1575 cm^{-1} /608/. Due to relaxation of the long wavelength Raman selection rule in graphitic materials composed of very small crystallites, the line broadens and a second broad band at 1355 cm^{-1} grows in. The spectrum resembles the phonon density of states of graphite /609/, behaviour characteristic of the amorphous phase. As the Raman features from contaminated silver (e.g. Fig. 50a) are very similar to those spectra, they are attributed to amorphous graphitic carbon deposits /607/. This conclusion was subsequently confirmed /610/ with XPS for films evaporated at 300 K in poor vacuum ($\approx 10^{-6}$ Torr), which showed intense Raman bands at 1380 cm^{-1} and 1590 cm^{-1}. The unusual strength of the graphitic carbon bands has been attributed to the extraordinary large Raman cross section of graphitic carbon (a factor of 50 larger than benzene) combined with an electromagnetic enhancement ($\approx 10^2$, /607/) due to excitation of surface plasmon resonances at the rough silver surface (in /607/, silver was deposited on 100 nm CaF_2, which provides a microscopically rough substrate).

The spectrum from the sputter-cleaned surface displayed in Fig. 50a is similarly interpreted. Here an additional contribution from subsurface carbon /25/ is possible. The origin of the bands at ≈ 800 cm^{-1}, 1162 cm^{-1}, and 1960 cm^{-1} is unclear. They might be due to bending and stretching modes of isolated C_x units, where x is of the order 2 - 4. As the dominating features in "impurity" spectra from coldly evaporated silver films (preceeding section) are due to adsorbed C_2 species (Fig. 49), it is conceivable, that some have clustered to larger units. The broad features at ≈ 1300 cm^{-1} and underneath ν_{C-C} of ethylene at 1586 cm^{-1} (Fig. 49) might indicate graphitic carbon nuclei on the surface.

To close we quote /607/: "The large Raman cross section of graphitic carbon suggests that Raman scattering may be a very useful technique for the observation of carbon contamination on metal and even semiconductor surfaces".

6.4 Summary

Carbon monoxide interacts only weakly with silver films at 120 K. As ν_{C-0} from SER spectra is identical to the frequency of the isolated molecule, the molecule is presumably only physisorbed. Interaction with gold or copper films is stronger as indicated by CO stretching frequencies at 2118 cm^{-1} (Au) and 2104 cm^{-1} (Cu) in agreement with IR data. Further studies are necessary to confirm present results and interpretations and to elaborate details of the bonding to the surface. This holds also for the interesting results from carbon monoxide on nickel. Carbonaceous and amorphous carbon deposits on silver, in quantities below the detection limit of other surface sensitive techniques such as AES, may lead to pronounced bands in SER spectra. This has to be taken into account when interpreting SER spectra from exposed films, since these species usually represent a major part of the trace impurities in the gas supply and background pressure of UHV systems. The very high sensitivity of SERS for amorphous carbon makes it a valuable tool for detection of carbon contaminants on metals.

7. Oxygen Exposure

The adsorption of oxygen and the possible subsequent oxidation of metal surfaces is an important process, for instance in oxidation reactions over metal catalysts like that of ammonia for nitric acid synthesis (see, e.g., /387/). Therefore many investigations employing various surface sensitive techniques have been concerned with oxygen-metal interactions. Silver has attracted particular attention, as this metal selectively catalyzes the epoxidation of ethylene to ethylene oxide /390 - 393/.

The free oxygen molecule is well characterized: the 0-0 stretching frequency (ν_{0-0}) is found at 1556 cm^{-1} /562/, and optical absorption is dominated by a strong band in the UV (Schumann-Runge band, $\approx$ 7 - 9 eV, /428,562,611/). With respect to adsorption on metals, the properties of metal-dioxygen complexes /474,612 - 614/ and of oxygen atoms bonded to one or several metal atoms /560,615/ are of special interest. The metal-oxygen stretching frequency of the latter is found around 950 cm^{-1} in inorganic compounds, when oxygen is linked to one metal atom (approximately double bond), and it is found around 600 cm^{-1} or less, when it is bonded to two or more metal atoms /560,616/. Matrix-isolated AgO /617/ and CuO /618/ display ν_{Me-O} at 499 cm^{-1} and 628 cm^{-1} in good agreement with gas phase data (485 cm^{-1} /619/ and 631 cm^{-1} /620/). The vibrational energies of dioxygen complexes depend on the oxidation state (i.e. bond order n) of the oxygen molecule. Most compounds belong to the two main classes, superoxo (n = 1,5) and peroxo complexes (n = 1). In either class the oxygen molecule may be bonded to one metal atom or may bridge two metal atoms /560,614/. Metal-dioxygen stretching frequencies are usually below 600 cm^{-1}, with ν_{Me-O_2} of the peroxide being somewhat larger than ν_{Me-O_2} of the corresponding superoxide (see, e.g., /621/). Silver superoxide isolated in a solid O_2/Ar matrix displays a single ν_{Ag-O_2} frequency at 440 cm^{-1} /622,623/ indicating a non-symmetric, bent Ag-0-0 structure (/623/; ν_{Me-O_2} from complexes of other metals from group Ib have not been reported). An approximately linear relationship exists between the bond order n and the stretching frequency as shown in Fig. 51 /614,621/. Dioxygen vibrational frequencies of the superoxo (1075 - 1195 cm^{-1}) and peroxo complexes (790 - 932 cm^{-1}) are well separated /614/. The frequencies of some noble metal superoxides in O_2/Ar matrices /622 - 625/ are indicated in Fig. 51.

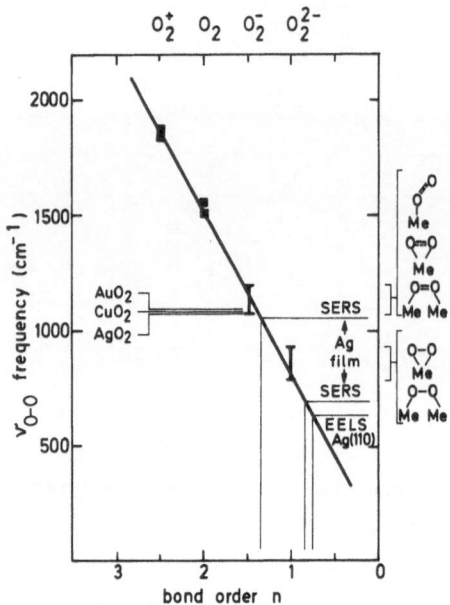

Fig. 51. Stretching frequency ν_{0-0} as a function of the bond order n of the oxygen-oxygen bonding (after /614,621/). Stretching frequencies of some matrix-isolated noble metal superoxides (IR studies /622 - 625/), of dioxygen on Ag(110) at 110 K (EELS studies /633,634/), and on Ag films at 120 K (SER studies /239,654/) are indicated. Possible structures of peroxo and superoxo complexes are schematically drawn at the right hand side [bond order: n = 1 (———); n = 1.5 (═══)]

Oxygen adsorbs dissociatively on many metal surfaces at or above room temperature (e.g. on W(100) /626/, Ni(100) /627,628/, Ru(001) /629/, Cu(100) /630/, Pt(111) /631, 632/, Ag(110) /633,634/, and on Fe(110) /635/). Besides physisorption of molecular oxygen on several metals at very low temperatures ($\lesssim$ 30 K, /636 - 639/), chemisorbed *molecular* oxygen species have been found only on Pt(111) /631,632/ and polycrystalline Pt /640/, on evaporated Ga films /638/, on Ag(110) /633,634,641/, and on polycrystalline gold foils /642/ (T $\approx$ 100 - 150 K; adsorption on low index faces of copper /643 - 645/ is treated below). Associatively adsorbed oxygen on gold foils exposed to $\gtrsim 10^4$ L O_2 at 100 K desorbs molecularly from the surface on warming, without leaving behind atomic oxygen /642/, which is a rather unique behaviour. Several studies are concerned with copper (/644,645/ and further references therein). Above room temperature, chemisorbed atomic oxygen is responsible for the single line in vibrational spectra at 395 cm^{-1} (Cu(110), /643/), at $\approx$ 300 cm^{-1} (Cu(100), /646/), and at 237 cm^{-1} (Cu(111), /647/; a shoulder at 403 cm^{-1} is attributed to oxygen atoms on defect sites). At elevated temperature ($\approx$ 450 K), slow diffusion of oxygen into the bulk is observed /646/. At the same time, Cu_2O islands are formed, which grow into bulk cuprous oxide films (/647/; Cu(111) in contact with 2 Torr of oxygen). Inconsistent results are reported for adsorption between 100 K and 300 K. Whereas atomic oxygen only is detected in EELS spectra from exposed Cu(110) /643/, UPS data suggest the coexistence of atomic and molecularly adsorbed singlet (!) oxygen on Cu(110) in this temperature range (/644/; Cu(100) and Cu(111) are reported to behave similarly /644,645/).

In general motivated with the industrial importance of the silver-catalyzed partial oxidation of ethylene, numerous studies have been devoted to oxygen adsorption on silver. Adsorption kinetics on supported and unsupported Ag (e.g. /648,649/) point to the existence of several adsorption states, whose identity is difficult to work out because of the heterogeneity of these surfaces. Below $\approx$ 170 K, oxygen is molecularly adsorbed on Ag(110) /633,634,641/ (ν_{0-0} = 630 cm^{-1}, ν_{Ag-O_2} = 240 cm^{-1}; /633/). It dissociates into adsorbed atoms, usually accompanied by desorption of part of the diatomic species, at T $\gtrsim$ 180 K (ν_{Ag-O} = 315 cm^{-1} /633/). Diffusion of adsorbed atoms leads to subsurface oxygen at T $\approx$ 450 K /633/. At T $\gtrsim$ 500 K atomic oxygen is desorbed. The very small value of ν_{0-0} indicates considerable weakening of the O-O bond upon adsorption (bond order $\approx$ 0.75, Fig. 51). This is explained with complete electron donation from the metal into the π^*-antibonding orbitals accompanied by further donation into the σ^*-antibonding orbitals /634/ or backdonation to the metal from the filled π-bonding orbital /633/. Oxygen adsorption on other low index faces of Ag (e.g. /650 - 652/) is less detailed investigated. A perfect Ag(111) surface seems to be inert to oxygen exposures below 10^{-3} Torr at room temperature, adsorption occurs only on defect sites /651,652/. Ag(100) behaves presumably similarly /651/, available data are rather scarce. The importance of defects for oxygen adsorption on Ag has recently been demonstrated with UPS studies on coldly evaporated films /277/. A surface defect concentration of 10 or 20 per cent has been estimated. It was concluded, that oxygen adsorbs associatively only on these defect sites (T = 140 K). Only physisorbed oxygen is present on silver films evaporated and exposed at 20 K /353,639,653/. This species may be considered as a precursor of the more strongly bonded species observed at $\approx$ 100 K /639/.

Finally, we point to the absorption band at $\approx$ 355 nm in electron energy loss spectra from O_2 on silver films prepared at 20 K /263/. These were identified as metal-molecule charge transfer excitations localized at sites of microscopic roughness which might contribute to SERS from these systems /263/ (matrix-isolated AgO$_2$ absorbs at 275 nm /622/).

7.1 Silver Samples

Figure 52 displays Raman spectra from coldly evaporated silver films exposed to oxygen /239,654/. The upper two spectra are from samples, which were dosed by backfilling the UHV chamber through a variable leak valve (ion pump and gauge running during exposure). Four bands between 650 cm^{-1} and 1300 cm^{-1} are observed. All shift to smaller energy when $^{18}O_2$ is used. They are accompanied by a broad feature extending from $\approx$ 200 cm^{-1} to $\approx$ 500 cm^{-1} (peak at $\approx$ 250 cm^{-1}). The spectra suggest, that a variety of oxygen species and/or coadsorbed, *oxygen containing* impurities are present on the surface after exposure. To extract contributions from the latter, the experimental conditions were improved by using a nozzle beam doser directed onto the freshly prepared silver surface [(c) and (d) in Fig. 52]. To minimize gas conversion, the ion pump was valved off and the system pumped by a turbo molecular pump during exposure [(d) in Fig. 52]. The spectrum recorded under the most appropriate experimental conditions (d) shows only two of the lines seen in other spectra [at 1053 cm^{-1} and at 697 cm^{-1}; another weak band might be hidden in the tail of the Rayleigh line at $\approx$ 250 cm^{-1}; some spectra taken under similar conditions as for (d) display more pronounced low energy bands between 200 cm^{-1} and 500 cm^{-1} similar to (c) (see also Fig. 53)].

A tentative assignment of the observed Raman features is given in Table 12. The most reproducible lines at 697 cm^{-1} and 1053 cm^{-1} are attributed to the O-O stretching ing vibrations of two different dioxygen species /654/ bonded to surface defect sites /277/. According to Fig. 51, they are identified as a superoxo-like species

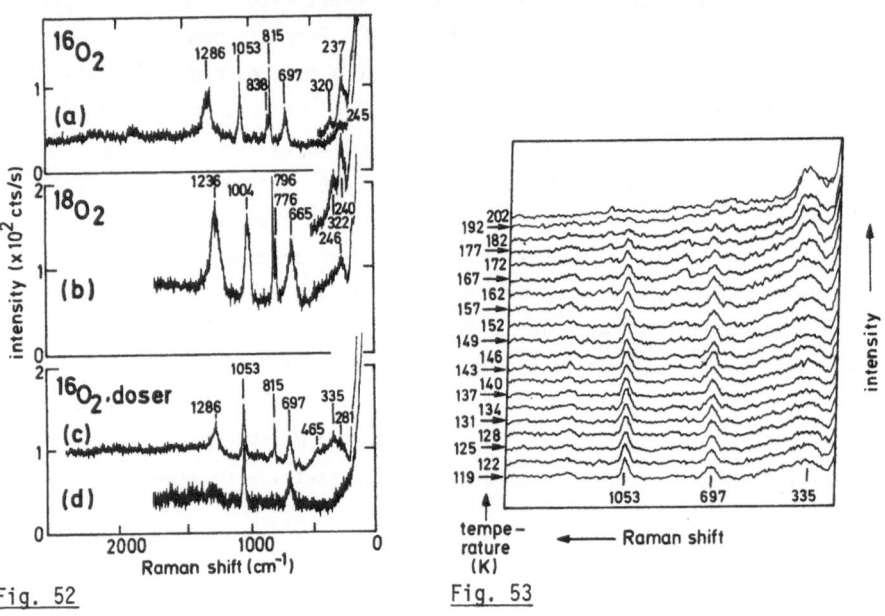

Fig. 52   Fig. 53

Fig. 52. SER spectra from coldly evaporated silver films exposed to oxygen.
(a): sample exposed to 10^3 L of $^{16}O_2$ by back-filling the chamber, ion pump running, T = 120 K; (b): like (a), but 300 L of $^{18}O_2$; inset shows low energy features after annealing to 200 K; (c): sample exposed by using a beam doser (actual dose unknown), ion pump running, T = 120 K; (d): like (c), but ion pump valved off (chamber evacuated with turbo molecular pump). All spectra have been taken with 200 mW of 514.5 nm radiation and 4 cm^{-1} bandpass. After /239,654/

Fig. 53. Annealing of SER spectra from oxygen exposed silver film. Exposure as for (d) in Fig. 52. Spectra have been taken with an OMA system (200 mW, 514.5 nm). After /654/

Table 12. Vibrational frequencies of various oxygen species. (a): SER study of oxygen exposed silver films [after /239,654/; values marked by an asterisk are calculated with the isotope shift expected for free molecules ($\nu^{18}/\nu^{16} = 0.943$)]; (b): EELS study of oxygen on Ag(110) (after /633,634/); (c): IR study of matrix-isolated AgO /617/, AgO$_2$ /622,623/, and AgO$_3$ /617/. Vibrational frequencies of carbon dioxide and carbonate species on Ag(110) are displayed for comparison [(d), EELS study /661/; here the silver surface was covered with 10 per cent of a monolayer of atomic oxygen at 170 K and subsequently exposed to CO$_2$ at 100 K]

Mode Species	(a) SERS $^{16}O_2$	(a) SERS $^{18}O_2$	(b) EELS Ag(110)	(c) Matrix isolated AgO[1], AgO$_2$[2], AgO$_3$[3]	(d) Carbonate Spectra Ag(110)	(d) Mode Species
ν_{Ag-O_2}	~ 200	~ 200	240	445/498[2]		
ν_{Ag-O}	~ 500	~ 500	315	499[1]	270 (310)	ν_{Ag-O}
ν_{0-0}, O_2^{2-}			630			
0-0 (bridge)	697	665 (657)*			660	δ_{O-C-O}, CO_2
ν_{asym}, O_3^-	815	776		791/798[3]		
? O–O–O ?	838	796			850	π_{CO_3}, CO_3
ν_{0-0}, O_2^-	1053	1004 (993)*		1079/1084[2]	1050	ν_{C-O}, CO_3
O–O ?	1286	1236			1280	δ_{O-C-O}, CO_2
					1360	ν^{s}_{O-C-O}, CO_3
					1390	ν^{s}_{O-C-O}, CO_2
					2350	ν^{a}_{O-C-O}, CO_2

(n $\approx$ 1.4, 1053 cm^{-1}) and a peroxo-like species (n $\approx$ 0.8, 697 cm^{-1}). The assignment is supported by vibrational data from silver complexes /622,623/ and from exposed Ag(110) surfaces (/633,634/; presumably due to the absence of suitable defect sites on well prepared Ag(110), no superoxide species is observed). Furthermore, the lines display the correct isotope shift (Table 12) and the expected annealing behaviour. As described above, dioxygen on Ag(110) dissociates into atomic oxygen at $\approx$ 180 K /633,634/. ν_{O-O} bands in Raman spectra from exposed silver films start to disappear at $\approx$ 170 K (Fig. 53). Simultaneously, the broad low energy feature attributed to ν_{Ag-O_2} and ν_{Ag-O} (Table 12) gains intensity, mainly on its high energy side, which indicates formation of atomic oxygen. The shape of this band suggests coexistence of atomic and molecular oxygen already at 120 K on the sample used for the annealing experiment (Fig. 53). Other samples [(a),(b),(d) of Fig. 52] show less evidence of adsorbed atomic oxygen at this temperature. Here the intensity of the band is concentrated around $\approx$ 250 cm^{-1}, which corresponds to ν_{Ag-O_2}. The Raman signal from atomic oxygen is lost between 240 K and 260 K /655/, where coldly evaporated silver films become SERS inactive (see Chapt. 2). Since oxygen atoms do not desorb in this temperature range /633,634/, the observed Raman features are surface enhanced (this is also obvious from the intensity of the dioxygen lines, Fig. 52, which originate from less than a monolayer of adsorbed molecules /277/).

The remaining SER lines (Fig. 52) are presumably due to molecules, which are reactively formed during oxygen exposure in the UHV chamber and/or at the silver surface. The doublet at 815/818 cm^{-1} is most pronounced, when the ion pump and gauge are running during exposure (Fig. 52, (a) and (b); it is even stronger, if the silver film is evaporated in $1 \cdot 10^{-5}$ Torr oxygen /656/). It seems likely, that ozone molecules are synthesized at hot filaments and in the ion pump. As matrix-isolated metal-ozonides display the asymmetrical oxygen stretching frequency at $\approx$ 800 cm^{-1} /617,618,657,658/, we assign the doublet tentatively to an adsorbed ozonide species (with presumably site-split $\nu_{a_{O-O}}$). The observed isotope shift is in good agreement with the corresponding value of AgO_3 (44 cm^{-1}, /617/). It is, however, unclear, why the bending mode ($\approx$ 600 cm^{-1}, /618,658/) and particularly $\nu_{s_{O-O}}$ ($\approx$ 1020 cm^{-1}, /658/), which is very strong in ozonide Raman spectra, are absent in SER spectra (SERS related change of selection rules as discussed in, e.g., /67/?). Further experiments are necessary to clear the situation.

The species responsible for the band at 1286 cm^{-1} is unknown. As is evident from the $^{18}O_2$ spectrum (Fig. 52b), oxygen is certainly involved. Samples, which yield spectra as shown in Fig. 52a, display two bands at 240 cm^{-1} and 320 cm^{-1} after annealing to 200 K. Similar bands have previously been assigned to an adsorbed C_2 species (see Chapts. 5,6). This suggests the presence of carbonaceous contaminants on the exposed silver surface. Therefore the band at 1286 cm^{-1} could be caused by a carbonate species formed by reaction of adsorbed atomic oxygen with carbon dioxide from the ambient /659,660/. As molecular oxygen and carbonate can co-exist at silver surfaces /641/, adsorbed dioxygen species are not affected. Corresponding reactions have been studied on Ag(110) /661/. Vibrational frequencies of adsorbed CO_2, carbonate, and atomic oxygen from this investigation are listed in Table 12. As CO_2 features are irrelevant (CO_2 desorbs at $\approx$ 120 K /661/), an interpretation based on *monodentate* carbonate frequencies as observed in /661/ is impossible. This species has no vibrational mode around 1280 cm^{-1}, but on the other hand, a strong band around 1360 cm^{-1} which is not observed in SERS. Assuming the presence of bidentate or both carbonate species /659,662/ on the surface of our silver film would lift the former difficulty ($\nu_{aO-C-O} \approx 1280$ cm^{-1}, /661/), but would raise another question, since no band is observed in the 1450 - 1700 cm^{-1} range where $\nu_{C=O}$ is expected. It follows, that contributions from carbonate species to the spectra obtained under "improper" dosing conditions are unlikely. Any final conclusion would, however, be premature.

Only one further Raman vibrational study of oxygen adsorption on silver has been performed so far /586/. Two peaks at 260 cm^{-1} and 324 cm^{-1} have been observed after exposing a coldly evaporated silver surface (T $\approx$ 110 K) to $2 \cdot 10^3$ L of oxygen. These were assigned to atomic oxygen on two different sites. No trace of adsorbed dioxygen was found in this study.

7.2 Other

Coldly Evaporated Al Films. No "clean" Raman spectrum could be obtained from freshly prepared, coldly evaporated aluminum films /144/. Several bands between 500 cm^{-1} and 1200 cm^{-1} similar to SER lines from oxygen exposed silver films were observed. Further bands at 1810 cm^{-1}, 2110 cm^{-1}, and 3140 cm^{-1} suggest the presence of hydrocarbon and C_2 species on the surface (Chapt. 6). To examine the role of oxygen and to prevent contamination by other molecules, a film was deposited in $2 \cdot 10^{-8}$ Torr of oxygen. Only the bands at 560, 845, 930, and 1100 cm^{-1} similar to those from the nominally clean Al surface were detected. These lines were tentatively assigned to molecular oxygen, which is a rather preliminary conclusion as outlined /144/. More data from samples prepared under improved experimental conditions have to be collected before any final interpretation can be given.

Polydiacetylene Single Crystals. Polydiacetylene is a wide-gap semiconductor with low electrical conductivity /663/. Raman spectra from a freshly polymerized (100) face exposed to air display a band at 1521 cm^{-1} which has been attributed to adsorbed O_2 /164/ (in addition to lines of the polymer backbone). This band escaped detection in earlier Raman studies (e.g. /664/), because its intensity shows a narrow resonance at 2.39 eV as a function of incident laser photon energy (FWHM $\approx$ 0.01 eV). It was concluded /164/, that SERS is observed from adsorbed O_2 due to a well defined electronic transition with significant charge transfer from the polydiacetylene backbone to the O_2 molecule.

7.3 Summary

Two adsorbed dioxygen species were identified in SER spectra from silver films at 120 K: a superoxo- (O_2^-) and a peroxo-like (O_2^{2-}) species. These are bonded to defect sites /277/. At $\approx$ 170 K, both species dissociate to form adsorbed atomic oxygen. The relative intensities of oxygen related as well as of "impurity" SER features observed after oxygen exposure depend on the experimental design, for instance the dosing technique. Reactions with molecules from the ambient may also be important (atomic oxygen is rapidly consumed by, e.g., CO to form CO_2 /659/). The origin of the "impurity" lines is not clear at present. Species reactively formed at hot filaments or at the silver surface might be involved (e.g. ozone, carbonaceous molecules). SER studies of samples exposed to ozone and of surface reactions of atomic oxygen on Ag with, e.g., CO_2 and CO could help to answer open questions.

8. Water Adsorption

The bonding of water to metal surfaces is of fundamental importance in electrochemistry, catalysis, and corrosion. The nature and orientation of the adsorbed species, the adsorption site, and the interaction with the metal as well as between adsorbed water molecules are particularly interesting. As for SERS, it is remarkable that enhanced water signals are usually absent in spectra from electrodes, although a large fraction of the surface is covered by water molecules /665/. Special treatment, which probably leads to the formation of surface complexes involving halide ions and water molecules, is necessary to observe SER signals form water on electrodes /666 - 670/.

The non-linear, symmetric water molecule (symmetry C_{2v}, $\sphericalangle$ HOH = 104.5^0, dipole moment 1.84 D /671/) has three fundamentals ($2A_1 + 1B_1$), which all are Raman as well as infrared active /1 , 467/. Raman scattering from the bending (scissors) mode δ is very weak. The corresponding cross section is about three orders of magnitude smaller than that of the breathing mode of benzene /21/. Ice, lattice water, and aquo complexes exhibit, in addition to the three lines of the isolated molecule, bands below ≈ 900 cm^{-1} due to restricted rotations (librations) and restricted translations. Because of hydrogen bond interactions (e.g. /672/), similar features are also found in vibrational spectra from liquid water. Vibrational properties of water in various forms and environments are extensively discussed in several review articles and books /474,671,673 - 676/.

On most clean metal surfaces, water adsorbs associatively at temperatures below ≈ 150 K (e.g. /630,677 - 688/; dissociatively adsorbed water giving hydroxyl species and hydrogen has been found on Fe(100) at 130 K /689/ and on Cu(110) for T $\gtrsim$ 180 K /690/). Usually, the molecules are weakly bonded to the substrate via the lone-pair orbitals of the oxygen atom. Hydrogen bonded species have been reported for Pt(100) /677/ and Fe(100) /689/. As the tendency of water to form intermolecular hydrogen bonds is quite pronounced /671/, cluster formation and island growth is frequently observed, even at low temperature and exposure. Water *monomers* have so far been detected only on Ru(001) at 85 K /683/ and Cu(110) at 90 K /690/. Exposures exceeding several Langmuir lead to multilayer adsorption (T $\lesssim$ 150 K). Vibrational spectra are very similar to those of the natural forms of ice or vitreous ice. Desorption of

multilayer water is observed at ≈ 160 K, desorption of first layer water usually
at ≈ 180 K. Water reacts eagerly with pre-adsorbed atomic oxygen to form hydroxyl
species at temperatures above ≈ 150 K.

Water adsorbs without dissociation on both, the clean and oxygen covered Ag(110)
surface at 100 K /686,687/. Even for sub-monolayer coverage vibrational spectra closely resemble those of ice I_h, which suggests formation of three dimensional clusters
on the surface. Water desorbs at 170 K without reaction from the clean surface. Several reactions involving formation and recombination of hydroxyl groups are observed
on the oxygen pre-covered surface between ≈ 200 K and ≈ 320 K /686,687/.

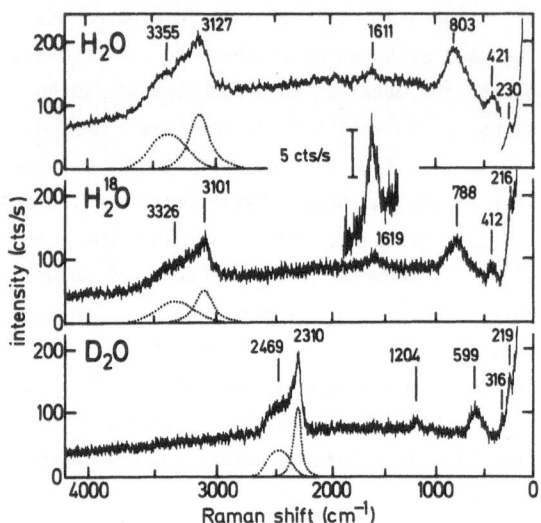

Fig. 54. SER spectra from coldly
evaporated silver films exposed
to 10 L of water. All spectra have
been taken with 200 mW of 514.5 nm
radiation and 6 cm^{-1} bandpass (different time constant for H_2O spectrum). After /691/

Figure 54 displays Raman spectra from coldly evaporated silver films exposed to
10 L of H_2O, $H_2^{18}O$, and D_2O, respectively /691/. In the region of the O-H stretching
vibrations (ν_s) a broad band between 3100 cm^{-1} and 3400 cm^{-1} is observed (between
2300 cm^{-1} and 2500 cm^{-1} for D_2O), which may be decomposed into two broad lines (see
also below). The relative small energies of 3355 cm^{-1} and 3127 cm^{-1} (compared to
3657 cm^{-1} (ν_1) and 3756 cm^{-1} (ν_3) for water vapour /671/) are characteristic of intermolecular hydrogen bonding (e.g. /672/). The weak peak at 1611 cm^{-1} (1204 cm^{-1}
for D_2O) is assigned to the bending mode δ. This feature is clearly resolved, if the
background intensity is suppressed in a high sensitivity scan (see inset for $H_2^{18}O$
in Fig. 54). Maxima below ≈ 900 cm^{-1} (below ≈ 700 cm^{-1} for D_2O) are attributed to
restricted rotations and translations of the water molecule.

Table 13 shows the assignment of the observed SER features in detail. It also displays vibrational data of solid water (amorphous ice I_v /692,693/ and hexagonal ice
I_h /694,695/) and EELS results from adsorbed water multilayers on Pt /677,678/ and

Table 13. Vibrational frequencies of ice I_v (a: after /693/; b: after /692/), ice I_h (c: after /695/; d: after /694/), and adsorbed water (e: on Ag(110), after /686/; f: on Pt(100), after /677/; g: on Pt(111), after /678/). Isotope values in parentheses are calculated /1/ from measured H_2O SER data, numbers in square brackets give the halfwidth (FWHM) of the observed bands. SER data after /691/

Vibrational Mode	Raman, IR				EELS		SERS		
	Ice I_v		Ice I_h		Pt(100), Pt(111) Ag(110)		Coldly evaporated Ag film		
	H_2O	D_2O	H_2O	D_2O	H_2O	D_2O	H_2O	D_2O	$H_2^{18}O$
stretching	3477^a	2476^a							
(ν_s)	3370^a	2476^a	3320^c	2489^c	3410^e		3355	2469	3326
					3330^f	2500^f		(2459)	(3335)
					3400^g	2530^g	[330]	[250]	[340]
	3220^a	2370^a	3210^c	2416^c					
	3112^a	2315^a	3085^c	2283^c			3127	2310	3101
								(2292)	(3109)
							[170]	[80]	[170]
bending	1660^b	1212^b	1650^d	1210^d	1660^e		1611	1204	1619
(scissors, δ)					1650^f	1220^f		(1195)	(1610)
					1625^g	1200^g	[120]	[90]	[120]
libration	802^b	600^b	840^d	640^d	740^e		803	599	788
					840^f	640^f		(599)	(798)
(rocking R_r,					700^g	(550 –	[240]	[150]	[230]
wagging R_w,						$700)^g$			
twisting R_t)							421	316	412
								(315)	(418)
							[60]	[60]	[75]
translation	213^b		225^c	217^c	200^e		230	219	216
(T_z)					240^f	240^f		(218)	(218)
					250^g	240^g			

Ag /686/. The interpretation is supported by isotope line frequencies calculated /1/ from measured SER H_2O data (last two columns in Table 13, values in parentheses). From simple moment of inertia considerations one estimates a ratio $\nu(H_2O)/\nu(D_2O) =$ 1.34 for the wagging libration R_w. This value agrees well with the observed isotopic shifts for both, the 803 cm^{-1} and the 421 cm^{-1} line (rocking and twisting librations have corresponding ratios of 1.39 and 1.41 /681/). Therefore an assignment to wagging vibrations of two distinct types of water admolecules or to intermolecularly coupled wagging librations (see, e.g., /675/) is more likely than an attribution to two different types of librations (as for instance proposed to explain some vibrational features of water on Ag(110) /686/). Correspondingly, the observation of (at least) two peaks in the O-H stretching region is probably due to a splitting of the O-H···O stretching frequency caused by intra- and intermolecular coupling /675/ and/or molecules located in different environments. The O-H stretching Raman band of amorphous ice I_V, which is very similar to the corresponding SER feature, is decomposed into *four* lines in /693/ (see Fig. 55a). These are attributed to three different types of molecules, i.e. molecules differently bonded to their neighbours. A strongly hydrogen-bonded species contributes two lines (ν_{sym} and $\nu_{antisym}$; $\nu_{s,a}^{sb}$ in Fig. 55a), a bent hydrogen-bonded and a weakly hydrogen-bonded species each one line (ν_{sym}; ν_s^{bb} and ν_s^{wb} in Fig. 55a). In addition, the first overtone of the bending mode δ enhanced by Fermi resonance with ν_s might be hidden in the O-H stretching band (e.g. /675/). Because of the breadth of the distribution of 2δ state in ice /675/ this contribution

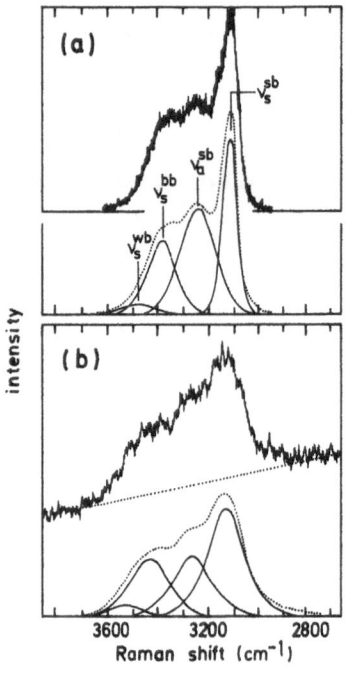

intensity

Raman shift (cm^{-1})

Fig. 55. (a) Decomposition of intramolecular stretching mode band of amorphous ice I_V as given in /693/ and (b) analogous decomposition of corresponding SER band of the spectrum displayed in Fig. 54

is, however, presumably small. Analogous to ice I_v, the O-H stretching band in SER spectra may be decomposed into four lines (Fig. 55b). This seems reasonable in view of the details of the actually measured band shape (Fig. 55b) and the much larger width of the high energy feature, when the band is decomposed into only two lines Fig. 54 and Table 13). We emphasize, however, that SER spectral features *alone* do not justify this decomposition, which therefore must be considered both tentative and speculative. On the basis of the present data, a more detailed discussion of the SER O-H stretching band to clear the open questions is not possible.

In summary, our SER vibrational data compare favourably to those of amorphous ice I_v and adsorbed multilayers on metals. Therefore, disordered water multilayers or islands of multilayers must be responsible for the observed spectra. This conclusion is confirmed when annealing the sample. The characteristic SER features disappear at ≈ 175 K indicating desorption of multilayer H_2O in agreement with recent results from Ag(110) /686/.

The Raman spectra are certainly enhanced, because only a few water layers contribute to the spectra. According to /677/, a water exposure of 10 L at 120 K results in the formation of 5 layers of ice on the sample. This value is probably slightly smaller in our experimental set-up, because a certain fraction of water molecules might be trapped at cold parts of the sample holder before reaching the Ag surface. No characteristic SER features could be detected for sub-monolayer coatings ($\lesssim 1$ L exposure). Water lines start to emerge from the background for exposures of several Langmuir and saturate in intensity for $\approx 10 - 20$ L. Additional condensation of further layers (100 L) does not increase the line intensity. In agreement with other results (see, e.g., Chapt. 4), this points to a short range enhancement mechanism for the investigated system.

Compared to SERS from other adsorbates on coldly evaporated silver films (e.g. pyridine or ethylene, Chapts. 4 , 5), water displays relatively small SER intensities. For the scissors mode δ this might be explained with the extremely small ordinary Raman cross section of this vibration /21/. It is, however, not clear, why the O-H stretching vibrations are so weakly pronounced. The low SER intensity of adsorbed water molecules explains, why water usually is not detected in SER spectra from silver electrodes. SER signals are simply overwhelmed by those from bulk water. Nevertheless, SER vibrational studies of adsorbed water are feasible as has been demonstrated. Further interesting problems are, for instance, water adsorption on Cu films and interaction of water with pre-adsorbed oxygen atoms.

9. Other Adsorbates

9.1 Diatomic Species

Nitrogen. SER spectra from nitrogen condensed on coldly evaporated silver films at 11 K exhibit a single line due to the nitrogen stretching vibration ν_{N-N} at 2326 cm^{-1} /133/, which is close to the corresponding gas phase value (2331 cm^{-1}, /562/). Compared to coadsorbed CO, an approximately 100 times smaller enhancement is estimated. This has been attributed to the fact, that silver is inert to nitrogen and therefore a chemical enhancement is not possible (CO is believed to form a bond, albeit a weak one; /133/ and Chapt. 6). The interpretation is corroborated by the absence of any pronounced first layer effect. On the other hand, a distinct first layer effect is observed for nitrogen on lithium. N_2 interacts stronger with coldly evaporated Li films as indicated by the downshift (19 cm^{-1} to 2307 cm^{-1}) and broadening of the ν_{N-N} SER band /133/. This is not surprising, since lithium and nitrogen are known to readily react under ordinary conditions.

Cyanide. So far, no SER studies of adsorbed cyanide have been performed under clean ultra high vacuum conditions. All published spectra were taken in air. The samples were mechanically abraded and immersed in KCN solution /90/ or electrochemically treated in dilute cyanide solutions /111,696,697/, and subsequently rinsed with destilled water and dried with nitrogen. Metal island films were exposed to a HCN atmosphere for a few minutes /91,137,145,698/. All procedures are assumed to result in roughly a monolayer of adsorbed cyanide. SER spectra from silver samples exhibit a strong line at $\approx$ 2140 cm^{-1} /90,91,696,697/, which is attributed to the stretching vibration ν_{C-N} (corresponding frequencies of silver cyanide complexes are listed in /87/; the observed SER frequency is almost identical with that of K[Ag(CN)$_2$]). A broad shoulder between 225 cm^{-1} and 260 cm^{-1} is assigned to ν_{Ag-CN-} /91/. Some spectra /91,697/ display more than one line in the region of the C-N stretching vibration, which has been attributed to different adsorbed species analogous to cyanide adsorption on silver electrodes (see, e.g., /87/). Moreover, reactions with molecules from the ambient (e.g. with oxygen /696/) and formation of new species are also possible. Indeed, such effects might be involved in the frequently observed fast decay of the

SER signal from cyanide on Ag in air (within $\approx$ 1 hour, /696/). SER spectra from cyanide on gold are also time dependent. The strong line at 2144 cm^{-1} (ν_{C-N}, /145/) converts into a broader and stable line at 2191 cm^{-1} within 20 min. It is speculated, that reaction of the adsorbed CN molecules with, for instance, water vapour forms another CN-containing complex at the surface. Cyanide on electrochemically processed Cu foils displays Raman spectra similar to those of solid CuCN (/140/; ν_{C-N} = 2174 cm^{-1} (2170 cm^{-1}), ν_{Cu-CN^-} = 319 cm^{-1} (319 cm^{-1}); values in parentheses are from ordinary spectra of CuCN powder). The enhanced signal is believed to originate from multi-layers of CuCN on copper metal (estimated enhancement: $\approx 10^3 - 10^4$, /140/).

In summary, cyanide on metals of group Ib gives rise to strong SER spectra. In-formation on, for instance, the chemical identity or the bonding geometry of the ad-sorbed species may be extracted from SER spectra, if the unknown and uncontrollable influence of the ambient is eliminated by performing experiments under ultra high vacuum conditions.

9.2 Azabenzenes

From the three azabenzenes studied by SERS, s-triazine $C_3H_3N_3$, pyrazine $C_4H_4N_2$, and pyridine C_5H_5N, the latter has been treated in detail in Chapt. 4. Like pyridine, the former two molecules display a very structured SER spectrum when condensed on coldly evaporated Ag films at 11 K /133/. Assignment of the lines has been accom-plished by comparison to bulk phase data. Only minor frequency shifts were observed. This indicates weak interaction with the silver surface, similar to pyridine. As in SER spectra from other adsorbates (Chapts. 4,5), a number of normally Raman-forbid-den bands are seen. The relative intensities of the spectral features change with time upon laser irradiation or sample heating to 40 K for all three azabenzenes, which is interpreted as arising from a surface molecular rearrangement /133/. Thermal or laser heating is assumed to provide the energy to overcome an activation barrier between two bonding geometries: the "flat" mode of bonding and the "standing up" ad-sorption geometry, where the latter is energetically more favourable at high cover-age /133/. Most of the freshly deposited adsorbate will, however, adopt the former mode of bonding, since the molecules approach the surface with an orientation favour-able to flat adsorption. For details of the argumentation the interested reader is referred to the original literature /133/. The picture discussed in /133/ is certain-ly a rather simplified approach to reality considering the high degree of disorder at the surface of coldly evaporated films (see Chapt. 3).

9.3 Pyridine Derivatives

Various n-pyridine carboxylic acids (isonicotinic acids) have been used for basic studies /44,62,92,294,552/ (for similar reasons as benzene derivatives, see Sect. 5.2.2). The molecules are usually adsorbed from solution or spin-deposited on SERS active island films of silver (or of gold /62,92/). Since most measurements are performed in air, contaminants from the atmosphere might be responsible for part of the observed features. Under "optimum" conditions /92/, some Raman bands from iso-nicotinic acids are $\approx 10^5$ times enhanced. Spectra from gold samples are weaker by a factor of about 100 than the corresponding spectra from Ag island films, and they are only observable for excitation wavelengths above 520 nm /62/. A detailed analysis of the SER spectral features has not been performed. It is remarkable, that 4-pyridine carboxylic acid on silver displays many distinct SER lines, whereas 2-pyridine carboxylic acid shows only some broad features /552/ (this is similar to n-amino-benzoic acids, see Sect. 5.2.2). The significance of this observation has still to be worked out.

9.4 Polymer and Langmuir-Blodgett Coatings

Polystyrene. Raman signals from polystyrene layers, which were spin deposited on holographic gratings overcoated with silver /46,47/ or gold /699/ and on silver island films /258/, are enhanced by excitation of surface plasmon resonances. Excitation of various guided light modes in sufficiently thick polystyrene layers via the grating surface (see, e.g., /42/) leads apparently also to an enhancement of the Raman signal /699/ (this phenomenon has first been treated in /29/). Only the strong 1000 cm^{-1} ring breathing mode has usually been employed in the studies. An analysis of the vibrational data has not been performed.

Cd-arachidate. Raman spectra have been measured for cadmium arachidate [Cd$(C_{19}H_{39}CO_2)_2$] monolayer assemblies /700/ on silver grating surfaces /46,701,702/, on silver island films /701/, and sandwiched between a grating surface and an island film /701/. Vibrational spectra were enhanced by the high intensity of plasmon surface polariton fields excited via the periodical corrugation of the surface. Spectral features in the C-H stretching region (2800 - 3050 cm^{-1}) have been used to obtain information on the structure of the films. Raman spectra from 11 layers (≈ 30 nm) on gratings resemble those of solid long chain fatty acids. They differ from those of layers on island samples or sandwiched between a silver island film and a grating, which are similar to spectra from disordered liquid samples. This has been attributed to conformational disorder in the layers in contact with silver islands, whereas assemblies on the relatively smooth shallow grating surface seem to be highly ordered /701/ (i.e. have regularly packed all trans alkyl chains). By comparison to corresponding infrared spectra it was concluded, that the Raman signals for samples with island-like topography come from a distorted minority species located in regions where electromagnetic fields are enhanced by, e.g., shape effects like the "lightning rod" effect (see Sect. 2.2). The fast saturation with coating thickness (at ≈ 2 nm) of SERS from polystyrene on Ag island films has been explained similarly /258/.

9.5 Dye Molecules

SER studies have been performed for several dye molecules on Ag island films or coldly evaporated Ag films (fluorescein isothiocyanate /92,294/, rhodamine 6G /92, 93, 548,549/, basic fuchsin /548,549/, metal-free phthalocyanine /703/, and diphenyloctatetraene and diphenylhexatetraene /704/). The observed enhancement factor is much smaller than for transparent molecules (e.g. pyridine) on the same surfaces /548, 549,704/. This has been attributed /548,549/ to the broadening of the luminescent molecular electronic level caused by metal-molecule interaction /229,705‑707/. Isolated dye molecules are subject to resonance or pre-resonance Raman scattering /169/. Upon adsorption on, for instance, island films (physisorption assumed /548,549/), the Raman intensity is further increased by the local field enhancement due to excitation of electromagnetic resonances. However, part of this electromagnetic enhancement is offset by the weakening of the molecular resonance due to the increased width of the absorbing state. This reduces SERS by two or three orders of magnitude /548, 549/, but still makes resonance Raman spectroscopy feasible in cases where it would otherwise be obscured by fluorescence (fluorescence is only weakly amplified or even attenuated by the presence of the silver island film /548,549/).

Vibrational analysis of the SER spectra has usually not been performed. In closing we mention the amplification of Raman features from α-copper phthalocyanine when the molecule is pressed into a silver disc /708/. It is conceivable that electromagnetic enhancement mechanisms are involved. Final conclusions are, however, not possible since the system is poorly characterized.

9.6 List of Systems Studied so Far

The list of SER studies from solid/gas interfaces compiles the work available to us till March 1983. Only papers are quoted, which provide information on spectral details and its interpretation, e.g. an assignment of SER bands in comparison to ordinary spectra. As this is not a sharp restriction, the list may contain blanks where it should not, and I apologize to those, whose work is not considered although they feel it should. A more complete list of publications, which "indicate or could be interpreted to involve some form of surface enhancement" of Raman signals, has recently been published elsewhere /709/.

Silver

Carbon	/90/	(O 300, air);	/25/	(O 300, UHV)
	/607/	(O 300, UHV)		
Oxygen	/586/	(CF 110, UHV);	/239/	(CF 120, UHV);
	/654/	(CF 120, UHV)		
Nitrogen	/133/	(CF 11, HV)		
Carbon monoxide	/61/	(CF 80, UHV);	/136/	(CF 11, HV);
	/586/	(CF 110, UHV);	/132/	(O 11, UHV);
	/370/	(CF 10, UHV);	/587/	(CF 120, UHV)
Cyanide	/90/	(O 300, air);	/696/	(O 300, air);
	/697/	(O 300, air)		
Water	/691/	(CF 120, UHV)		
Acetylene	/402/	(O 11, UHV);	/133/	(CF 11, HV);
	/523/	(CF 120, UHV);	/525/	(CF 120, UHV)
Ethylene	/135/	(CF 11, HV);	/458/	(CF 120, UHV);
	/402/	(O 11, UHV);	/133/	(CF 11, HV);
	/459/	(CF 120, UHV);	/460/	(CF 120 UHV)
Propylene	/135/	(CF 11, HV)		
Butene	/401/	(CF 11, HV)		
Benzene	/116/	(CF 11, HV);	/118/	(CF 11, HV);
	/133/	(CF 11, HV);	/50/	(GS 100, UHV)
Pyridine	/96/	(O 77, HV);	/99/	(CF 120, UHV);
	/266/	(IF 5, UHV);	/133/	(CF 11, HV);
	/108/	(CF 120, UHV);	/123/	(CF 120, UHV);
	/710/	(CF, IF 15, UHV);	/94/	(CF, IF 15, UHV);
	/369/	(CF 15, HV);	/358/	(CF 120, UHV)
Pyrazine	/133/	(CF 11, HV)		
s-Triazine	/133/	(CF 11, HV)		
Benzoic acid	/62/	(IF 300, air);	/92/	(IF 300, air)
Nitrobenzoic acid	/44/	(O, IF 300, air)		
Aminobenzoic acid	/44/	(O, IF 300, air)		

Isonicotinic acid	/62/ (IF 300, air);	/44/	(O, IF 300, air);
	/552/ (IF 300, air);	/92/	(IF 300, air)
Metal-free phthalocyanine	/703/ (IF 300, air)		
Cd-arachidate	/701/ (GS 300, air)		

Copper

Carbon monoxide	/588/ (CF 120, UHV)
Cyanide	/140/ (O 300, air)
Ethylene	/462/ (CF 120, UHV)
Pyridine	/123/ (CF 120, UHV); /369/ (CF 15, HV);
	/273/ (CF 120-300, UHV)

Gold

Carbon monoxide	/586/ (CF 110, UHV)
Cyanide	/145/ (IF 300, air)
Pyridine	/123/ (CF 120, UHV); /369/ (CF 15, HV)

Lithium

Nitrogen	/133/ (CF 11, HV)
Benzene	/133/ (CF 11, HV)

Sodium

Benzene	/142/ (CF 15, HV)

Aluminium

Oxygen	/144/ (CF 120, UHV)

CF: coldly evaporated film; IF: island film; GS: grating surface; O: other; the number following these abbrevations gives the temperature, at which the sample has been prepared and/or the measurements have been performed; UHV, HV, air means that the sample was prepared/investigated in ultra high vacuum, high vacuum, or air, respectively.

10. Selected Applications and Related Surface Enhanced Phenomena

10.1 Tribology

Silver island films of ≈ 5 nm mass thickness have been spin-coated with diphenyl disulfide from methanol solution /711/. Surface-enhanced Raman spectra from adsorbed molecules differ from conventional spectra of bulk samples (Fig. 56). The band at 542 cm^{-1} in the latter spectrum, which is assigned to the S-S stretching vibration ν_{S-S}, is absent in the former. This suggests that S-S scission occurs in diphenyl

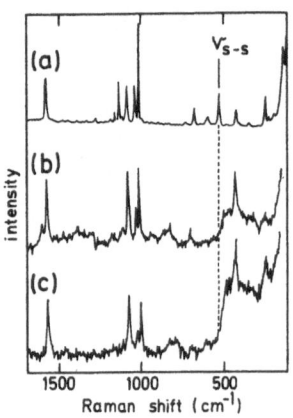

Fig. 56. Raman spectra of organic sulfides. (a): ordinary spectrum of bulk sample of diphenyl disulfide; (b): SERS from diphenyl disulfide on Ag island film; (c): SERS from phenyl mercaptan on Ag island film. SER spectra have been recorded with 15 mW of 514.5 nm radiation and 9 cm^{-1} bandpass. After /711/

disulfide upon adsorption on silver. The interpretation is corroborated by the fact, that the SER spectrum of the decomposition product closely resembles that of phenyl mercaptan on silver (curve c in Fig. 56). Similarly, diphenyl sulfide undergoes a surface reaction involving cleavage of the C-S bond which also leaves phenyl mercaptide on the surface. Analogous results have been obtained with the corresponding benzyl sulfides. It is remarkable, that S-S and C-S linkages in organic sulfides are easily cleaved on silver even under quite mild conditions (room temperature). As dibenzyl disulfide, for example, is a prototype of an "extreme pressure" additive

/712,713/, "the observation of adsorbed organo-sulfur monolayers by SERS offers a powerful means to elucidate the mechanism of action of antiwear additives and other lubricants" (quoted from /711/).

10.2 Catalysis

Raman scattering has been used to monitor the catalytic formation of SO_3^{2-} and SO_4^{2-} on the surface of Ag powder by exposure to sulfur dioxide gas /714/. Spectra from the fresh, finely divided powder in He atmosphere exhibit only two peaks in the spectral region between 400 cm^{-1} and 1200 cm^{-1} at $\approx$ 815 cm^{-1} and 1050 cm^{-1} (Fig. 57). These are believed to be associated with NO_2^- and NO_3^-, respectively /715/, since the Ag powder was reduced from $AgNO_3$. As the silver powder was in contact with oxygen from the ambient (air) during preparation, it is likely, that atomic oxygen is formed and adsorbed on its surface. Two new lines at 615 cm^{-1} and 925 cm^{-1} are seen after admittance of a 1-2 min pulse of SO_2 gas. These have been assigned to adsorbed SO_3^{2-}, catalytically formed on the Ag surface by reaction of the SO_2 gas with surface oxide. If the SO_2 exposed silver powder is kept in an atmosphere containing SO_2 and O_2 (1:5 ratio), the spectrum shown in Fig. 57c is measured at room temperature (note, that spectra b and c in Fig. 57 are almost identical). Slow heating of the sample in contact with SO_2/O_2 to 380 K causes pronounced changes: the two SO_3^{2-} peaks (615 cm^{-1}, 925 cm^{-1}) almost disappear, and a new peak develops at 962 cm^{-1}

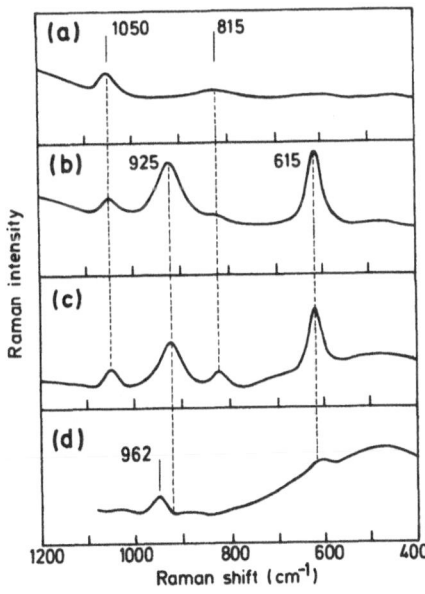

Fig. 57. SER spectra from finely divided Ag powder. (a): fresh powder in He atmosphere, T = 295 K; (b): like (a), but exposed to SO_2 gas for 1-2 min; (c): like (b), but kept in a SO_2/O_2 (ratio 1 to 5) containing atmosphere after initial exposure; (d): like (c), but slowly heated to 380 K. Spectra have been taken with 514.5 nm or 488 nm radiation and 4 cm^{-1} bandpass. After /714/

(Fig. 57d). It is concluded, that part of the SO_3^{2-} is oxidized to SO_4^{2-} yielding the new line, while the remainder of the SO_3^{2-} is thermally desorbed. Upon recooling to room temperature, the spectrum shown in Fig. 57d does not change. Furthermore, the spectrum displayed in Fig. 57c cannot be restored by exposing the powder once more to SO_2 at 295 K. This implies blockage or destruction of the sites active in catalytic formation of SO_3^{2-} /714/. As SO_2 is known as catalyst poison, which for instance blocks the oxygen adsorption centers important for ethylene oxidation /716/, further studies on this system seem to be worthwile. In a second paper /715/, the catalytic formation of $AgNO_2$ and $AgNO_3$ on the oxygen-contaminated Ag surface exposed to NO and NO_2 gas pulses has been reported.

It was assumed in /714/ that the observed Raman features were surface enhanced. This assumption seems reasonable, since electromagnetic resonances may well be excited in the particles of the Ag powder. Whether and to what extend other mechanisms contribute, as well as the magnitude of the overall enhancement, is as yet unclear. Regardless of these open questions, the observations presented in /714,715/ and in /161/ (CO/H_2 on heated Ni particles) provide some encouragement for Raman scattering as a potential diagnostic technique in catalytic investigations.

10.3 Other Surface Enhanced Processes

Fluorescence. The fluorescence of a molecule is usually quenched when it is placed close to a metal surface /229/, since additional nonradiative decay channels are opened by the metal /705,717,718/. If the molecule is placed near a suitably roughened metal surface (e.g. near an Ag island film), the quenching may be offset by the enhancement of the incident and emitted electromagnetic fields by surface plasmon resonances /92,294,296,548,549,707,719/. This effect is optimized, when the metal electromagnetic resonance coincides with the dye adsorption /296/. Compared to SERS, only moderate enhancement is observed (usually $\lesssim 10$ /296,707,719/). For given substrate, the magnitude of the enhancement depends on the quantum yield of the molecular fluorescence (/548,549/; Ag island film). Molecules with very low quantum efficiency display the greatest enhancement. This may be useful for fluorescence spectroscopy. For dyes with high quantum efficiency the enhancement may *not* be sufficient to balance the quenching mentioned above. In this case a *decrease* of the fluorescence is observed, even for Ag island films which exhibit strong SERS /548,549/.

Photochemistry. Analogous to fluorescence, the photochemical yield from molecules on rough surfaces depends on the competition between the surface enhanced absorption and surface enhanced damping processes /720/. Two extreme cases may be distinguished. Firstly, the chemical process following absorption of a photon is very fast, as for instance in direct photodissociation. In this case, the enhanced photochemistry is essentially determined by the enhanced absorption. Secondly, the chemical reaction requires accumulation of a certain energy corresponding to a few photons (e.g. multiphoton dissociation). Then the ability of the molecule to store energy becomes important. For molecules close to a *small* metal sphere, which roughly models a surface protrusion or an island film, it was recently estimated /720/, that substantial enhancements of photochemical yields in either case are possible. Unlike SERS, optimal performance is not given by molecules directly attached to the surface. Because of the different distance dependence of energy transfer and damping processes, an optimal molecule-surface separation exists, which depends on the parameters of the system (threshold for the reaction, lifetime of the activated molecule, shape and intensity of the exciting pulse /720/).

Experimental studies on photochemistry of molecules on silver island films /721/ and sputter-roughened Ag single crystal surfaces /722/ have recently been performed. The latter paper presents evidence for surface-enhanced photofragmentation of adsorbed pyridine, pyrazine, and benzaldehyde (but not for benzene; UV excitation). The exact nature of the excitation and fragmentation process is unclear, multiphoton absorption leading to ionization and decomposition is a likely mechanism /722/.

Infrared Absorption. Infrared reflection absorption spectroscopy of molecules on metal surfaces usually suffers from its low signal levels ($\approx 0.1\%$; only for strong absorbers like carbon monoxide up to 10% absorption are measured /583/). If infrared surface electromagnetic waves at the metal surface are excited in a suitable ATR arrangement, the sensitivity is increased by the substantial build-up of the electromagnetic field at the interface /723 - 725/. Two Mn-stearate Langmuir-Blodgett layers on silver led to easily detectable signals in the 1200 - 1300 cm^{-1} frequency region. Detection of even a sub-monolayer is possible as inferred from the magnitude of the absorption peaks /723/. Similarly, infrared absorption from molecules on island films may be enhanced by excitation of (infrared) collective electron resonances in the metal particles /726,727/. Spectra from a monolayer of stearic acid on an Ag island film of 5 nm thickness evaporated onto a Ge prism are shown in Fig. 58 /727/. The enhancement is clearly seen from a comparison of the bottom spectrum to the top spectra, but there remain some problems in interpreting the observations. The absence of any absorption for s-polarized radiation is difficult to understand within the frame of collective electron resonances and is, moreover, in conflict with the results of /726/. More experiments are needed to clear the situation.

Nonlinear Phenomena. Nonlinear optical effects are conveniently described by a power series expansion of the relationship between the polarization and the electric field (see, e.g., /728/). As the various higher order polarization terms in the power series expansion depend on the local electric field strength - each in its own characteristic way - nonlinear optical phenomena on metal surfaces may be enhanced by excitation of electromagnetic resonances similar to SERS /729,730/. Such effects

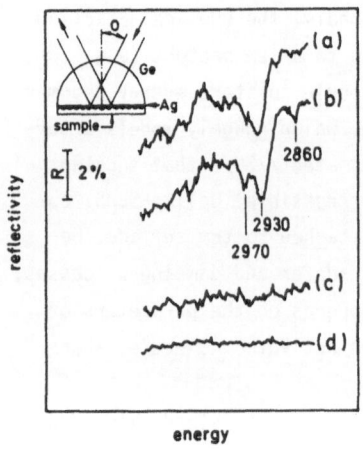

Fig. 58. Infrared absorption spectra of a monolayer of stearic acid on a silver island film
(5 nm mass thickness) evaporated onto a Ge prism.
(a): $\Theta = 16^0$, p polarized incident radiation;
(b): $\Theta = 20^0$, p polarized incident radiation;
(c): $\Theta = 16^0$, s polarized incident radiation;
(d): $\Theta = 20^0$, p polarized incident radiation,
without silver film. Inset shows experimental
arrangement. After /727/

have been investigated experimentally and/or theoretically for second harmonic generation /729-736/, coherent anti Stokes Raman scattering /737,738/, Raman gain
spectroscopy /111/, hyper Raman scattering /739/, and two photon fluorescence /740/.
As an example, we present some results from second harmonic generation. Figure 59
shows the spectral distribution of the non-linear signal from a silver surface which
had been electrochemically roughened (activated) in the same way as samples used
for SER studies /733/. The second harmonic signal appears as a sharp peak at the
second harmonic frequency (532 nm; the broad background is believed to be luminescence /733/). Compared to a smooth evaporated Ag film, the signal is roughly four
orders of magnitude enhanced, which corresponds to a local field enhancement of $\approx$ 20
/733/. Second harmonic generation from silver island films yields maximum intensity
for a certain mass thickness of the films (Fig. 60a; /734/), i.e. for a certain island size and distribution. As is anticipated from an electromagnetic model involving
localized surface plasmons /734/, the peak is observed in between the maxima of the
local field enhancement factors for the fundamental and the second harmonic frequency (Fig. 60b).

Since second harmonic signals originate from the first one or two layers of metal
atoms at the surface, it is clear that this technique is surface sensitive and may
be employed to study sub-monolayer amounts of adsorbates. Such investigations have
already been performed for pyridine and pyrazine on activated silver electrodes
/735,741/ (the latter is especially interesting, since second harmonic generation
is forbidden for centrosymmetric systems; upon adsorption the inversion symmetry obviously breaks down indicating appreciable interaction of the molecule with the substrate /741/). Other interesting results are reported in /111,739,740/ and especially
/738/, where an enhancement of the coherent anti Stokes Raman signal from a molecule
close to a silver particle of 21 (!) orders of magnitude under favourable conditions
has been calculated. The main advantage of the nonlinear optical techniques over

122

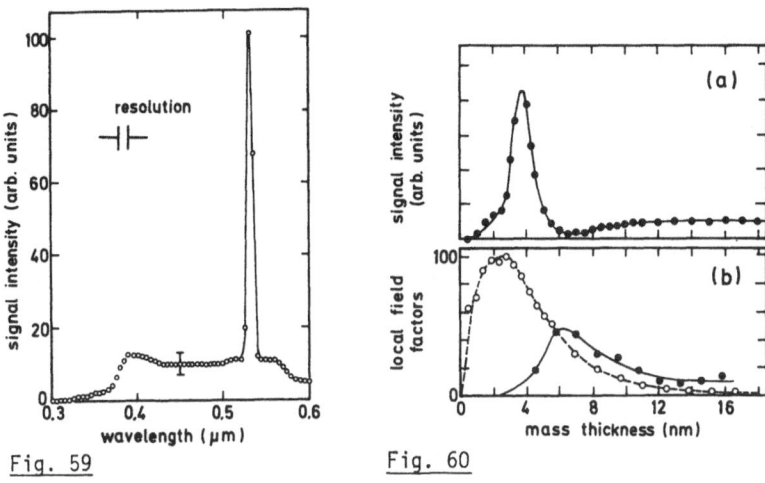

Fig. 59.

Fig. 60.

Fig. 59. Spectral distribution of nonlinear signal from a roughened silver surface. Excitation with ≈ 7 mJ pulses (10 ns) of 1064 nm radiation from a Q-switched Nd:YAG laser focused to 5 mm diam. spots (≈ 10 pulses/s). After /733/

Fig. 60. (a) Second harmonic intensity and (b) local field enhancement factors at 1064 nm (filled-in circles) and at 532 nm (open circles) from silver island film as a function of mass thickness. Excitation with 1.2 W average power of 1064 nm radiation from a mode locked cw Nd:YAG laser focused onto a 22 μm diam. spot. After /734/

other surface techniques including IRAS and SERS is the much better spectral resolution.

So far, enhanced nonlinear optical phenomena are only treated within a purely electromagnetic frame. Similar to SERS, this might be only one side of the coin. It is conceivable, that the various nonlinear polarizabilities and polarizability derivatives might experience adsorption induced changes analogous to the linear term. Future treatments have to address also this question.

For the sake of completeness, we finally mention another recent application of SERS /742/. Resonantly excited desorption of pyridine from silver island films by infrared laser absorption /743/ has been studied by following the SER signal from pyridine in the frequency region of the breathing modes. It was shown, that physisorbed pyridine is selectively desorbed by this technique.

11. Summary and Outlook

In *broad* terms, the phenomenon of surface enhanced Raman scattering seems to be understood. Several processes of different weight in different systems generally contribute to the overall enhancement. Electromagnetic mechanisms due to the excitation of surface plasmon type resonances are in principle understood. Reasonable quantitative description of actual systems is in many cases, however, still marginal. Weak resonance Raman scattering due to adsorption induced charge transfer excitations seems to be important for chemical effects. Here the *details* are unclear. Further insight in this process requires information on the local electronic structure of the adsorbate/adsorbent system, i.e. the chemisorption bond. As this changes with, for instance, adsorbate, adsorbent, and adsorption site, it is difficult to establish general rules for the chemical contribution to SERS. Pyridine, for example, experiences strong chemical enhancement only when bonded to certain defect sites on silver whereas these particular sites play a less pronounced role in SERS from ethylene on silver.

Strong or strongly enhanced Raman signals from adsorbed molecules are only expected (and observed) when several enhancement mechanisms work simultaneously. This limits the applicability of SERS to metals of high reflectivity with appropriately prepared surface, i.e. with suitable surface roughness. Weak or weakly enhanced (electromagnetic) Raman signals may be obtained from molecules on the relatively smooth surfaces of high reflectivity metals in an ATR or grating geometry. So far, there are no convincing experiments which unambiguously demonstrate a significant enhancement for molecules on a metal of low reflectivity where electromagnetic effects can be neglected.

These limitations are reflected by the experimental results from solid/gas interfaces published so far. Most SER studies have been dedicated to adsorption on coldly evaporated metal films (mainly group Ib and especially silver) which apparently combine several advantageous properties in a unique way. SER spectra of a variety of molecules on these surfaces have been recorded and used to obtain information on the bonding of the adsorbate to the surface. Unknown "impurity" species were identified in the spectra, and formation of new species on exposed surfaces upon annealing was traced by SERS. A detailed interpretation of spectral features is often rendered difficult by the fact that SERS active surfaces in general are poorly defined on an atomic scale. However, part of this disadvantage in comparison to, e.g., EELS from single crystal surfaces is offset by the high resolution of SERS which allows to discriminate a variety of adsorption states on the surface. Therefore SERS is about to establish itself as a surface analytical tool which can provide valuable information on molecule-substrate interaction, despite the fact that there are some unanswered questions concerning the details of the enhancement mechanism.

According to the specific advantages of surface Raman scattering in comparison to other techniques which probe vibrational properties of adsorbed molecules, there are several paths research could follow in the future.

Firstly, one should utilize the unique features of the noble metals of group Ib (coldly evaporated films): SER adsorption studies on silver and copper should be extended. The results presented in this review demonstrate both, the high sensitivity of SERS for organic (and other) deposits and the detailed information extractible from the spectra. As both, silver and copper, are important oxidation catalysts, SER adsorption studies of suitable molecules (e.g. ethylene oxide, methanol, formaldehyde) on clean and oxygen pre-dosed surfaces are extremely interesting. In addition, surface Raman studies on actual silver or copper catalysts seem to be feasible as might be inferred from the SER study of SO_2 oxidation over silver powder. Surface enhanced Raman investigations can contribute significantly to a better understanding of adsorption on metals of group Ib (the same holds presumably for the alkali metals).

Secondly, there may be a future in unenhanced Raman scattering from adsorbates on smooth surfaces, if the experimental conditions can further be improved. It has already been shown that sub-monolayers of strong scatterers can be detected on single crystal surfaces (*no* restriction to only certain metals). It seems worthwile to apply the advanced experimental technique of optical multichannel detection also to high area catalysts like supported and unsupported metals, where even better performance might be obtained.

Thirdly, the local field enhancement due to excitation of electromagnetic resonances in silver particles may be used to study other, catalytically relevant surfaces. In /744/, silver island films have been overcoated with thin aluminum layers (1.5 - 5 nm) which were subsequently completely oxidized. Organic molecules deposited on the aluminum oxide surface display strongly enhanced Raman scattering. Silicon oxide or mixed silicon / aluminum oxide surfaces may be prepared correspondingly. If appropriately overcoated with small metal clusters, these samples may also be useful for SER studies of supported metal catalysts. As several metals deposit on silver without interdiffusion into the bulk at room temperature, the metal overlayer may be brought directly onto the silver island /745/. Surface enhanced Raman scattering from cyanide on gold overlayers (0.1 - 1 nm) has already been observed, investigations of catalytically more important metals like nickel are in progress /745/.

Raman spectroscopy, enhanced or ordinary, will certainly consolidate its place within the community of surface analytical tools which probe vibrational properties of adsorbates at solid/gas interfaces. As SER is particularly suited for investigations of systems resembling the structure of real catalysts (roughened surfaces, island films), it may help to establish a link between adsorption and surface reaction studies on the perfectly characterized single crystal surfaces and the poorly defined catalyst surfaces.

Appendix:
Recent Developments and Results

Chapters 1-11 consider information available to the author in spring 1983. Here we briefly summarize recent results and update the list of references to spring 1984.

Introduction. Infrared vibrational spectroscopy of surface species is discussed in two recent review articles /746,747/, various new theoretical and experimental aspects of SERS are treated in /748,749/. The optimization of the experimental conditions for Raman scattering from ultra thin films in a three-boundary ATR arrangement (excitation of plasmon surface polaritons or guided light modes) is described in /750/. Further investigations of adsorbates by using unenhanced Raman scattering /751-754/ demonstrate the possible utility of this technique for surface studies.

Fundamentals of Surface Enhanced Raman Scattering. Several further substrate materials have been shown to display enhanced Raman scattering /755-758/:

i) A weak enhancement of ≈ 10 has been found for $Pt(CN)_4^{2-}$ ions on platinum colloids of 1.6 nm average diameter /755/ (excitation wavelength: 400 nm). The measured enhancement factor is consistent with Lorenz-Mie calculations. Nevertheless, possible contributions of other enhancement mechanisms could not be excluded.

ii) The enhancement of the Raman signal from disordered carbon on a rough, etched surface of lead telluride (enhancement factor $\approx 10^3$) has tentatively been attributed to nonvertical interband transitions in PbTe involved in the relevant intermediate states /756/. The spectrum is superposed on a broad continuum which is believed to originate from electronic excitations and/or vibrational modes intrinsic to adsorbed carbon. Similarly, the background continuum (or at least part of it) accompanying many SER spectra from metals may be due to carbon impurities rather than to an intrinsic property of the metal substrate. The role of carbon impurities in the corresponding SER experiments should carefully be re-examined.

iii) An enhancement factor of $\approx 10^4$ has been estimated for SERS from pyridine on a roughened β palladium hydride electrode /757/. As the electronic, and hence the optical, properties of β PdH are similar to those of silver /759,760/, the results might be explained analogous to corresponding observations from silver surfaces.

iv) Silicon microstructures exhibit enhanced Raman scattering from Si phonon modes /758/. The enhancement factor is $\approx 10^2$ for Si spheres of ≈ 0.1 μm diameter. The effect has been attributed to low order structure (Mie) resonances. This interpretation is backed by recent theoretical estimations of the electromagnetic field enhancement for dielectric spheres /761/ and an ordered two-dimensional array of spheres /762/ (the latter is particularly interesting since an appreciable effect is calculated even for ordinary dielectrics with small refractive index).

Concerning Raman scattering from smooth single crystal surfaces it has been shown, that the first layer enhancement is $\lesssim 4$ /763/ corroborating the statement given in Sect. 2.1. It follows, that image field effects are negligible for smooth surfaces (see also Sect. 2.3).

The electromagnetic theory of surface enhanced spectroscopy has recently been reviewed comprehensively /764/. Several further papers treat special configurations within an electromagnetic approach: (i) ATR arrangement to excite *long-range* plasmon surface polaritons /765,766/, (ii) optical gratings /767 - 771/ and bigratings /772/, (iii) set of deep narrow grooves corresponding to a rectangularly shaped grating /773/, (iv) small metallic sphere above a plane metallic substrate /774, 775/, and (v) isolated metallic particle (electrodynamic treatment of large silver spheroids /776/, contribution of surface scattering to plasmon resonance width /777/, effect of dielectric overlayers /778/). A coupled two-oscillator model is used in /779/ to calculate the Raman scattering from a molecule on an arbitrarily shaped metal surface and applied to the case of a metallic sphere. The important case of a randomly rough surface is treated with a nonperturbative approach in /780/.

The role of electromagnetic resonances in surface enhanced Raman scattering has experimentally been investigated for coldly evaporated Ag films /781 - 783/ and island films /784/. Evidence is presented, that the surface of the former is highly porous (intercrystallite gaps with ≥ 2 nm depth) and molecules *within* the cavities are subject to an electromagnetic enhancement ($\approx 3 \cdot 10^4$) due to cavity resonances /781/. This conclusion is doubted in /783,785/. For island films, the magnitude of the plasmon resonance contribution to SERS has been determined to be $\lesssim 10^3$ /784/. It was concluded, that other mechanisms provide an additional enhancement of $\approx 10^3$ for the investigated system.

The chemical contribution to SERS is further discussed in /786 - 801/. For the ground-state charge transfer model /224/ an enhancement factor of $10^2 - 10^3$ is estimated /787/. It is pointed out, that many characteristic SER features (e.g. specifity to adsorption site, appearance of forbidden bands) can be understood within this model. It is remarkable, that the model predicts a flat excitation profile. Due to participation of excitonic or interband excitations in the scattering process, Raman signals from molecules adsorbed on semiconductor surfaces may be enhanced /788,789/. Experimental evidence for this effect has been presented /756, 790 - 792/. The role of charge transfer excitations in SERS (excited-state charge transfer model /236 - 238/) is discussed in various recent experimental and theoretical papers /793 - 801/. Particularly noteworthy is the statement given in /801/, that advanced models of SERS should comprise both aspects, ground-state *and* excited-state charge transfer (these are termed "vibrational driven hopping" and "coherent tunneling" charge transfer in /801/). The assumption, that the chemical contribution to SERS is particularly strong at or restricted to defect sites (concept of SERS active sites), is supported by various recent results from mainly electrode surfaces /802 - 808/, although controversial views still exist /802,809,810/.

Time dependent Hartree-Fock calculations have been used to determine the Raman enhancement of Li_n - H_2 clusters /811/. The calculated polarizability derivative enhancement of 10^3 - 10^4 may be partitioned into three terms, an electromagnetic part, a term arising from the modulation of the cluster metal orbitals by the vibration of the adsorbate, and a part involving charge transfer between metal and adsorbate.

Laser beam induced photodecomposition effects and their impact on the interpretation of SER features from mainly electrode surfaces are discussed in /812 - 818/.

Experimental. Optical properties, structure and surface roughness of coldly evaporated films and their relation to SERS are discussed in /819 - 824/. The influence of the size dependence of the dielectric function on the optical absorption of island films is investigated in /825/. Electronic and vibrational features of matrix-isolated clusters and small particles (noble metals, silicon) are studied in /826 - 832/.

Pyridine Adsorption. Angle resolved UPS measurements from pyridine on Pd(111) at room temperature suggest bonding to the surface through both the nitrogen atom and the ring plane /833/. Therefore an adsorption geometry with the ring plane tilted with respect to the surface plane is proposed, which is similar to the high coverage phase of pyridine on Ag(111) (see /20/ and Chap. 4).

A detailed SER study of pyridine on copper (colloids) is presented in /834/. SER signals from pyridine on β palladium hydride (electrode) are discussed in /757/. Continuing earlier work /153,154,157/, Raman scattering from pyridine adsorbed to various vacuum evaporated metal films (room temperature, 10^{-5} Torr) as well as to semiconductor surfaces is studied in /835/. An enhancement factor of $\approx 10^4$ has been estimated for pyridine on Ag, Pd, and Ni (strongest effect for Ni). The conclusions are in conflict with those of other groups (see, e.g., /272/ and Sect. 3.2).

Hydrocarbon Adsorption. The bonding and surface chemistry of hydrocarbons and hydrocarbon fragments on metals has recently been reviewed /836 - 839/.

Numerous new investigations treat adsorption and surface reactions of ethylene /840 - 849/, propylene /850 - 852/, and acetylene /841,842,853 - 856/ on metal surfaces (mainly transition metals of group VIII, but also silver /844,847,852/). Decomposition of adsorbed unsaturated hydrocarbons into hydrogen-impoverished species has frequently been studied /837,840,848,851,854/. These investigations are interesting with respect to the results discussed in Sects. 5.1.2 and 5.1.4, where annealing induced changes of SER spectra from coldly evaporated silver films exposed to ethylene and acetylene have been explained with formation of such species. The role of oxygen pre-exposure for adsorption of ethylene on silver is investigated in /844, 847/. An enhanced interaction was observed even at 77 K /847/.

SER spectra of ethylene on coldly evaporated silver films obtained by a Russian group /857/ are in line with those displayed in Sect. 5.1.2. A detailed SER study of aminobenzoic acid on silver (colloids) is presented in /858/.

Carbon Monoxide Exposure and Carbonaceous Deposits. The continuing interest in carbon monoxide interaction with metal surfaces is reflected by the great number of papers still published in this field. State of the art overviews on bonding, vibrational features, and surface reactions are given in /746,859,860/. Vibrational spectra and their relation to the structure of CO adlayers on various single crystal metal surfaces are discussed in /861 - 869/, coverage dependent shifts and changes of the line shape of vibrational bands are treated in /746,866,870 - 872/. Studies of carbon monoxide interaction with copper /873 - 878/ and silver /844,878,879/ accentuate the role of defect sites /875 - 877/ and lateral interaction of the vibrating molecules /875,877,879/ for the interpretation of the observed band frequencies and shapes. In agreement with SER data (Sect. 6.1), a downshift of the CO stretching frequency from 2140 cm^{-1} to 2110 cm^{-1} with increasing coverage on silver has been measured and attributed to dipole-dipole coupling between the adsorbed molecules /879/ (IR study of matrix-isolated Ag clusters).

Due to its importance for heterogeneous catalysis, thermally (or otherwise) activated fragmentation of adsorbed hydrocarbons or carbon monoxide and formation of carbonaceous layers on metal surfaces has been the subject of several recent studies (e.g. /837,854,880 - 882/). Even at low temperature (140 K) cyclotrimerization of acetylene to benzene has been observed on Pd(111) /854/.

The dependence of Raman spectral features on structure and microtexture of carbon films is discussed in /883/.

Oxygen Exposure. Interaction of oxygen with group VIII metal surfaces is treated in several recent papers /860 - 862,884 - 891/. Chemisorbed molecular species have been found to co-exist with atomic oxygen also on Pd(100) /884/ and Rh(100) /890/ (T ≈ 100 K). The role of defect sites (steps, kinks) for oxygen adsorption is investigated in /861,862,889,892/.

Several studies are concerned with oxygen adsorption on polycrystalline silver films /847,893/ or foils /894,895/ and silver single crystal surfaces /844,896 - 901/. Uncertainty in interpreting the enormous downshift of ν_{O-O} of molecular oxygen on Ag(110) compared to the gas phase value still persists /896,897/. A tentative interpretation in terms of singlet oxygen analogous to the system O_2/Cu(110) /644/ has been presented /896/. A very small sticking coefficient has been found for oxygen adsorption on Ag(111) /898/ corroborating earlier results /651,652/. Whereas no molecularly adsorbed oxygen was found on Ag(111) at 100 K /846/, chemisorbed O_2^{m-} has been detected on cesium- and potassium-dosed silver foils even at room temperature /894/ as well as on polycrystalline foils of Ni, Cu, Ag, and Pt at 80 K /895/. In fair agreement with EELS data from Ag(110), vibrational bands at $\approx$ 240 cm^{-1}, 314 cm^{-1}, and 630 cm^{-1} have been reported for Ag foils /895/.

With respect to Raman studies of coldly evaporated Al films exposed to oxygen (Sect. 7.2), the IR investigation of matrix reaction products of oxygen and ozone with aluminum atoms /902/ is interesting.

Water Adsorption. A detailed theoretical analysis of the O-H stretching band of liquid water and ice is given in /903/. Several recent investigations treat water adsorption and interaction with adsorbed oxygen atoms on various metal surfaces /904 - 911/. Formation of OH species has been reported to start at temperatures as low as 80 K /910/ (oxygen pre-coated Ag(100) surface). Water monomers have been observed on Cu(100) and Pd(100) at 10 K /911/ (for small exposures $\lesssim$ 0.4 L). They start to cluster when warming the sample to only 20 K.

The necessity of working with electrolytes of high salt concentration to observe SERS from water on silver electrodes /666,667/ is attributed to an increased density of active sites at high salt concentration as well as to a more complete hydrogen bond disruption in the Helmholtz layer /912/ (i.e. a higher density of water-halide complexes). In /913/ it is pointed out, that the absence of SER features from *normal* water in electrode spectra may be explained with purely electromagnetic arguments.

Other Adsorbates. Recently, SER spectra from nitrogen and carbon dioxide on coldly evaporated silver films deposited at 120 K and cooled down to $\approx$ 40 K have been observed /914/. Nitrogen displays a single vibrational line at 2321 cm^{-1} close to the gas phase value in agreement with earlier results from samples deposited at 11 K /133/ (on coldly evaporated copper films a band at 2282 cm^{-1} has been reported /783/). Four bands are observed after CO_2 exposure at 653 cm^{-1} (ν_1, δ_{O-C-O}), 1278/1371 cm^{-1} [probably Fermi resonance of $2\nu_1$ with ν_2 (ν_{sO-C-O})], and 2343 cm^{-1} (ν_3, ν_{aO-C-O}) respectively. The measured frequencies are close to EELS data from Ag(110) /661/ and the gas phase values /1/.

The importance of chemical effects for SERS is revealed by a study of Raman scattering and luminescence intensities from crystal violet on smooth and roughened films of Ag and Au /915/. It is emphasized, that Raman scattering by adsorbed molecules should be viewed as scattering by the entire adsorbate/substrate complex. This point is also evident from a SER study of meso-tetraphenylporphine on Ag in a layered structure /916/.

Finally, enhanced Raman scattering from crystal vibrational modes of an antimony film laid down on a silver island substrate has been reported /917/ (enhancement factor: $\approx$ 20).

Selected Applications and Related Surface Enhanced Phenomena. SERS has been used (i) to study formation of NO_2^- and NO_3^- on the pre-oxidized surface of Ag powder cata- lysts after exposure to NO and NO_2/N_2O_4 /918/, (ii) to determine the extent of charge transfer between metal surfaces and chemisorbed molecules from shifts of vibrational frequencies /919/, (iii) to characterize silver-modified n-GaAs(100) photoelectrodes /920/, (iv) to obtain structural information on adsorbed amphiphilic molecules which are used to alter wettability and surface tensions at liquid/solid interfaces /921/, and (v) to investigate the oxide layer on Ag and Cu smoke parti- cles /922/.

For basic fuchsin molecules placed at varying separation from a Ag island film by means of a SiO_x spacer layer, maximum luminescence enhancement of 200 has been observed for a separation of $\approx$ 2.5 nm /923,924/. The result is explained with the competition between local electromagnetic field enhancement and loss of excitation by radiationless energy transfer to the metal. Picosecond fluorescence relaxation measurements of rhodamine 6G on silver island films are described in /925/.

Enhanced infrared absorption from monolayer species on silver films in an ATR arrangement is investigated in /926/. An interesting re-interpretation of early IR transmission experiments on coldly evaporated copper films exposed to CO /574/ is given in /927/. The observed Fano-type line shape of the C-O stretching band has been assigned to interference of continuous e-h-pair excitations in the metal (excited via surface defects by the IR radiation) and the discrete vibrational excitations. For surfaces with a high density of defects such as coldly evaporated copper films, the indirect excitation of the vibrational mode via infrared e-h-pair excitations has been postulated to be stronger than the direct photon-vibration interaction for CO on smooth surfaces.

Recent experimental results and theoretical developments concerning surface en- hanced nonlinear optical processes are described in /928 - 936/. Besides those ef- fects mentioned in Sect. 10.3, enhanced four-wave mixing has found some interest /931,933/.

References

1   G. Herzberg: *Infrared and Raman Spectra of Polyatomic Molecules* (Van Nostrand, New York 1945)
2   E.B. Wilson, Jr., P.C. Decius, P.C. Cross: *Molecular Vibrations* (McGraw-Hill, New York 1955)
3   H.C. Allen, Jr., P.C. Cross: *Molecular Vib-Rotors* (Wiley, New York 1963)
4   N.B. Colthup, L.H. Daly, S.E. Wiberley: *Introduction to Infrared and Raman Spectroscopy* (Academic Press, New York 1975)
5   R.J.H. Clark, R.E. Hester (eds.): *Advances in Infrared and Raman Spectroscopy* (Heyden, London 1978)
6   L.J. Bellamy: *The Infrared Spectra of Complex Molecules* (Chapman and Hall, London 1980)
7   N.L. Alpert, W.E. Kaiser, H.A. Szymanski: *IR, Theory and Practice of Infrared Spectroscopy* (Plenum, New York 1970)
8   A. Anderson (ed.): *The Raman Effect* (Dekker, New York 1971)
9   S.C. Brown (ed.): *Electron-Molecule Scattering* (Wiley, New York 1979)
10 K. Takayanagi, N. Oda (eds.): Proc. of the XIth Int. Conf. on the Phys. of Electronic and Atomic Collisions (North-Holland, Amsterdam 1979)
11 J.W. White: "Neutron Scattering Spectroscopy in Relation to Electromagnetic Methods", in *Molecular Spectroscopy*, ed. by P. Hepple (Institute of Petroleum, London 1972) p. 199
12 S.W. Lovesey, T. Springer (eds.): *Dynamics of Solids and Liquids by Neutron Scattering*, Topics in Current Physics, Vol. 3 (Springer, Berlin, Heidelberg, New York 1977)
13 L.H. Little: *Infrared Spectra of Adsorbed Species* (Academic Press, New York 1966)
14 M.L. Hair: *Infrared Spectroscopy in Surface Chemistry* (Dekker, New York 1967)
15 H. Ibach (ed.): *Electron Spectroscopy for Surface Analysis*, Topics in Current Physics, Vol. 4 (Springer, Berlin, Heidelberg, New York 1977)
16 H. Ibach, D.L. Mills: *Electron Energy Loss Spectroscopy and Surface Vibrations* (Academic Press, New York 1982)
17 R.F. Willis (ed.): *Vibrational Spectroscopy of Adsorbates*, Springer Series in Chemical Physics, Vol. 17 (Springer, Berlin, Heidelberg, New York 1980)
18 T. Wolfram (ed.): *Inelastic Electron Tunneling Spectroscopy*, Springer Series in Solid State Sciences, Vol. 4 (Springer, Berlin, Heidelberg, New York 1978)
19 R.P. Van Duyne: "Laser Excitation of Raman Scattering from Adsorbed Molecules on Electrode Surfaces", in Chemical and Biochemical Applications of Lasers, ed. by C.B. Moore, Vol. 4 (Academic Press, New York 1979) p. 101
20 J.E. Demuth, K. Christmann, P.N. Sanda: Chem. Phys. Lett. *76*, 201 (1980)
21 H.W. Schrötter, H.W. Klöckner: "Raman Scattering Cross Sections in Gases and Liquids", in *Raman Spectroscopy of Gases and Liquids*, ed. by A. Weber, Topics in Current Physics, Vol. 11 (Springer, Berlin, Heidelberg, New York 1979) p. 123
22 J.G. Skinner, W.G. Nilsen: J. Opt. Soc. Am. *58*, 113 (1968)
23 P.J. Hendra, E.J. Loader: Trans. Faraday Soc. *67*, 82 (1971)
24 R.G. Greenler, T.L. Slager: Spectrochim. Acta *29A*, 193 (1973)
25 I. Pockrand, A. Otto: Applic. Surf. Sci. *6*, 362 (1980)

26 Chih-Cong Chou, C.E. Reed, J.C. Hemminger, S. Ushioda: J. Electron Spectrosc. Relat. Phenom. *29*, 401 (1983)
27 A. Campion, J.K. Brown, V.M. Grizzle: Surf. Sci. *115*, L 153 (1982)
28 A. Campion: J. Electron Spectrosc. Relat. Phenom. *29*, 397 (1983)
29 Y. Levy, C. Imbert, J. Cipriani, S. Racine, R. Dupeyrat: Opt. Commun. *11*, 66 (1974)
30 J. Cipriani, S. Racine, R. Dupeyrat, H. Hasmonay, M. Dupeyrat, Y. Levy, C. Imbert: Opt. Commun. *11*, 70 (1974)
31 J. Cipriani, H. Hasmonay, Y. Levy, S. Racine, M. Dupeyrat, R. Dupeyrat, C. Imbert: Japan. J. Appl. Phys. *14*, Suppl. 1 - 4, 93 (1975)
32 M. Menetrier, R. Dupeyrat, Y. Levy, G. Imbert: Opt. Commun. *21*, 162 (1977)
33 H. Raether: "Surface Plasma Oscillations and Their Applications", in Physics of Thin Films, ed. by G. Hass, M.H. Francombe, R.W. Hoffmann, Vol. 9 (Academic Press, New York 1977) p. 145
34 Y.J. Chen, W.P. Chen, E. Burstein: Phys. Rev. Lett. *36*, 1207 (1976)
35 W.H. Weber, G.W. Ford: Opt. Lett. *6*, 122 (1981)
36 K. Sakoda, K. Ohtaka, E. Hanamura: Sol. State Commun. *41*, 393 (1982)
37 J.R. Kirtley, S.S. Jha, J.C. Tsang: Sol. State Commun. *35*, 509 (1980)
38 S.S. Jha, J.R. Kirtley, J.C. Tsang: "Raman Intensities from Molecules Adsorbed on Rough Metal Surfaces", in Proc. VIIth Int. Conf. on Raman Spectroscopy, Ottawa, 1980; ed. by W.E. Murphy (North - Holland, Amsterdam 1980) p. 356
39 S.S. Jha, J.R. Kirtley, J.C. Tsang: Phys. Rev. *B22*, 3973 (1980)
40 H. Numata: J. Phys. Soc. Jap. *51*, 2575 (1982)
41 N. Garcia: Opt. Commun. *45*, 307 (1983)
42 I. Pockrand: J. Phys. D: Appl. Phys. *9*, 2423 (1976)
43 B. Pettinger, A. Tadjeddine, D.M. Kolb: Chem. Phys. Lett. *66*, 544 (1979)
44 R. Dornhaus, R.E. Benner, R.K. Chang, I. Chabay: Surf. Sci. *101*, 367 (1980)
45 H.W.K. Tom, C.K. Chen, A.R.B. de Castro, Y.R. Shen: Sol. State Commun. *41*, 249 (1982)
46 A. Girlando, J.G. Gordon II, D. Heitmann, M.R. Philpott, H. Seki, J.D. Swalen: Surf. Sci. *101*, 417 (1980)
47 A. Girlando, M.R. Philpott, D. Heitmann, J.D. Swalen, R. Santo: J. Chem. Phys. *72*, 5187 (1980)
48 P.N. Sanda, J.M. Warlaumont, J.E. Demuth, J.C. Tsang, K. Christmann, J.A. Bradley: Phys. Rev. Lett. *45*, 1519 (1980)
49 S. Ushioda, Y. Sasaki: Phys. Rev. *B27*, 1401 (1983)
50 J.C. Tsang, J.R. Kirtley, T.N. Theis: J. Chem. Phys. *77*, 641 (1982)
51 J.C. Tsang, J.R. Kirtley, J.A. Bradley: Phys. Rev. Lett. *43*, 772 (1979)
52 J.C. Tsang, J.R. Kirtley, T.N. Theis: Sol. State Commun. *35*, 667 (1980)
53 J.R. Kirtley, T.N. Theis, J.C. Tsang: Appl. Phys. Lett. *37*, 435 (1980)
54 J.R. Kirtley, J.C. Tsang, T.N. Theis, S.S. Jha: "Surface Enhanced Raman Scattering from Tunnel Junction Structures", in Proc. VIIth Int. Conf. on Raman Spectroscopy, Ottawa, 1980; ed. by W.E. Murphy (North - Holland, Amsterdam 1980) p. 386
55 D.L. Jeanmaire, R.P. Van Duyne: J. Electroanal. Chem. *84*, 1 (1977)
56 M.G. Albrecht, J.A. Creighton: J. Amer. Chem. Soc. *99*, 5215 (1977)
57 M. Fleischmann, P.J. Hendra, A.J. McQuillan: Chem. Phys. Lett. *26*, 163 (1974)
58 A.J. McQuillan, P.J. Hendra, M. Fleischmann: J. Electroanal. Chem. *65*, 933 (1975)
59 P.J. Hendra, M. Fleischmann: "Raman Spectroscopy at Surfaces", in *Topics in Surface Chemistry*, ed. by E. Kay, P.S. Bagus (Plenum, New York 1978) p. 373
60 J.A. Creighton, C.G. Blatchford, M.G. Grant: J. Chem. Soc. Faraday II *75*, 790 (1979)
61 T.H. Wood, M.V. Klein: J. Vac. Sci. Technol. *16*, 459 (1979)
62 E. Burstein, C.Y. Chen, S. Lundquist: "Giant Raman Scattering by Molecules Adsorbed on Metals: An Overview", in *Light Scattering in Solids*, ed. by J.L. Birman, H.Z. Cummins, K.K. Rebane (Plenum, New York 1979) p. 479
63 T.E. Furtak, J. Reyes: Surf. Sci. *93*, 351 (1980)
64 A. Otto: Applic. Surf. Sci. *6*, 309 (1980)
65 H. Ueby, S. Ichimura, H. Yamada: Surf. Sci. *119*, 433 (1982)
66 H. Metiu: "Surface Enhanced Spectroscopy", Progr. Surf. Sci., to be published

67 A. Otto: "Surface Enhanced Raman Scattering: "Classical" and "Chemical" Origins", in *Light Scattering in Solids IV*, ed. by M. Cardona, G. Güntherodt, Topics in Applied Physics, Vol. 54 (Springer, Berlin, Heidelberg, New York 1984) p. 289

68 K. Arya, R. Zeyher: "Theory of Surface Enhanced Raman Scattering" in *Light Scattering in Solids IV*, ed. by M. Cardona, G. Güntherodt, Topics in Applied Physics, Vol. 54 (Springer, Berlin, Heidelberg, New York 1984) p. 419

69 R.K. Chang, T.E. Furtak (eds.): *Surface Enhanced Raman Scattering* (Plenum, New York 1982)

70 J.A. Creighton: "Raman Spectroscopy of Adsorbates at Metal Surfaces", in *Vibrational Spectroscopy of Adsorbates*, ed. by R.F. Willis, Springer Series in Chemical Physics, Vol. 15 (Springer, Berlin, Heidelberg, New York 1980) p. 145

71 R.P. Van Duyne: J. Physique *38*, C5 - 239 (1977)

72 H. Yamada: Appl. Spectr. Rev. *17*, 227 (1981)

73 A. Otto: "Surface Enhanced Raman Scattering", in *Dynamics of Gas - Surface Interaction*, ed. by G. Benedek, U. Valbusa, Springer Series in Chemical Physics, Vol. 21 (Springer, Berlin, Heidelberg, New York 1982) p. 186

74 R. Dornhaus: "Surface Enhanced Raman Spectroscopy", in Festkörperprobleme, ed. by P. Grosse, Vol. XXII (Vieweg, Braunschweig 1982) p. 201

75 R.L. Birke, J.R. Lombardi: "The Influence of Surface Features on Surface Enhanced Raman Scattering", in Advances of Laser Spectroscopy, ed. by B.R. Garetz, J.R. Lombardi, Vol. 1 (Heyden, London 1982) p. 143

76 V.V. Marinyuk, R.M. Lazorenko-Manevich, Ya.M. Kolotyrkin: "Anomalously Intense Raman Scattering from Molecules Adsorbed on Metals" in *Advances in Physical Chemistry*, ed. by Ya.M. Kolotyrkin (MIR Publishers, Moscow 1982) p. 148

77 R. Dornhaus: to be published

78 W. Krasser, H. Ervens, A. Fadini, A.J. Renouprez: J. Raman Spectrosc. *9*, 80 (1980)

79 J.M. Stencel, E.B. Bradley: J. Raman Spectrosc. *8*, 203 (1979)

80 P.J. Hendra, J.R. Horder, E.J. Loader: J. Chem. Soc. (A), 1766 (1971)

81 H. Jeziorowski, H. Knözinger: Chem. Phys. Lett. *43*, 37 (1976)

82 W. Krasser, A. Ranade, E. Koglin: J. Raman Spectrosc. *6*, 209 (1977)

83 R.P. Cooney, M. Fleischmann, P.J. Hendra: J. C. S. Chem. Comm., 235 (1977)

84 R.P. Cooney, E.S. Reid, P.J. Hendra, M. Fleischmann: J. Amer. Chem. Soc. *99*, 2002 (1977)

85 R.P. Cooney, G. Curthoys, Nguyen The Tam: Adv. Cat. *24*, 293 (1975)

86 T.A. Egerton, A.H. Hardin: Catal. Rev. Sci. Eng. *11*, 71 (1975)

87 J. Billmann, G. Kovacs, A. Otto: Surf. Sci. *92*, 153 (1980)

88 B. Pettinger, U. Wenning: Chem. Phys. Lett. *56*, 253 (1978)

89 S.G. Schulz, M. Janik-Czachor, R.P. Van Duyne: Surf. Sci. *104*, 419 (1981)

90 A. Otto: Surf. Sci. *75*, L 392 (1978)

91 J.G. Bergmann, D.S. Chemla, P.F. Liao, A.M. Glass, A. Pinczuk, R.M. Hart, D.H. Olson: Opt. Lett. *6*, 33 (1981)

92 G. Ritchie, C.Y. Chen: "Raman Scattering and Luminescence by Molecules Adsorbed at Metal Island Films", in /69/, p. 361

93 D.A. Weitz, S. Garoff, T.J. Gramila: Opt. Lett. *7*, 166 (1982)

94 H. Seki: J. Chem. Phys. *76*, 4412 (1982)

95 M. Kerker, O. Siiman, L.A. Bumm, D.-S. Wang: Appl. Optics *19*, 3253; 4137 (1980)

96 R.R. Smardzewski, R.J. Colton, J.S. Murday: Chem. Phys. Lett. *68*, 53 (1979)

97 G.L. Eesley: Phys. Lett. *81A*, 193 (1981)

98 J.E. Rowe, C.V. Shank, D.A. Zwemer, C.A. Murray: Phys. Rev. Lett. *44*, 1770 (1980)

99 I. Pockrand, A. Otto: Sol. State Commun. *35*, 861 (1980)

100 I. Pockrand, A. Otto: Sol. State Commun. *38*, 1159 (1981)

101 M. Udagawa, Chih-Cong Chou, J.C. Hemminger, S. Ushioda: Phys. Rev. *B23*, 6843 (1981)

102 H. Metiu: "A Survey of Recent Theoretical Work", in /69/, p. 1

103 R.L. Birke, J.R. Lombardi, J.I. Gersten: Phys. Rev. Lett. *43*, 71 (1979)

104 C.Y. Chen, E. Burstein, S. Lundquist: Sol. State Commun. *32*, 63 (1979)

105 A. Otto, J. Timper, J. Billmann, G. Kovacs, I. Pockrand: Surf. Sci. *92*, L 55 (1980)

106 J. Timper, J. Billmann, A. Otto, I. Pockrand: Surf. Sci. *101*, 348 (1980)
107 A. Otto, J. Timper, J. Billmann, I. Pockrand: Phys. Rev. Lett. *45*, 46 (1980)
108 I. Pockrand, A. Otto: Sol. State Commun. *37*, 109 (1981)
109 I. Pockrand: unpublished
110 J.P. Heritage, J.G. Bergman: "Picosecond Raman Gain Studies of Molecular Vibrations on a Surface", in *Light Scattering in Solids*, ed. by J.L. Birman, H.Z. Cummins, K.K. Rebane (Plenum, New York 1979) p. 167
111 J.P. Heritage, J.G. Bergman, A. Pinczuk, J.M. Worlock: Chem. Phys. Lett. *67*, 229 (1979)
112 A. Otto: Surf. Sci. *92*, 145 (1980)
113 E. Burstein, C.Y. Shen: "Raman Scattering by Molecules Adsorbed at Metal Surfaces. The Role of Surface Roughness", in Proc. VIIth Int. Conf. on Raman Spectroscopy, Ottawa, 1980; ed. by W.E. Murphy (North-Holland, Amsterdam 1980) p. 346
114 G.R. Erdheim, R.L. Birke, J.R. Lombardi: Chem. Phys. Lett. *69*, 495 (1980)
115 R. Dornhaus, M.B. Long, R.E. Benner, R.K. Chang: Surf. Sci. *93*, 240 (1980)
116 M. Moskovits, D.P. DiLella: J. Chem. Phys. *73*, 6068 (1980)
117 J.K. Sass, H. Neff, M. Moskovits, S. Holloway: J. Phys. Chem. *85*, 621 (1981)
118 D.P. DiLella, R.H. Lipson, P. McBreen, M. Moskovits: J. Vac. Sci. Technol. *18*, 453 (1981)
119 I. Pockrand, J. Billmann, A. Otto: J. Chem. Phys. *78*, 6384 (1983)
120 B. Pettinger, U. Wenning, H. Wetzel: Surf. Sci. *101*, 409 (1980)
121 B. Pettinger, H. Wetzel: Ber. Bunsenges. Phys. Chem. *85*, 473 (1981)
122 B. Pettinger, H. Wetzel: "Organic and Inorganic Species at Ag, Cu, and Au Electrodes", in /69/, p. 293
123 I. Pockrand: Chem. Phys. Lett. *85*, 37 (1982)
124 B. Pettinger: Chem. Phys. Lett. *78*, 404 (1981)
125 J.A. Creighton, M.G. Albrecht, R.E. Hester, J.A.D. Matthew: Chem. Phys. Lett. *55*, 55 (1978)
126 B. Pettinger, U. Wenning, D.M. Kolb: Ber. Bunsenges. Phys. Chem. *82*, 1326 (1978)
127 C.G. Blatchford, J.R. Campbell, J.A. Creighton: Surf. Sci. *108*, 411 (1981)
128 H. Wetzel, H. Gerischer: Chem. Phys. Lett. *79*, 460 (1981)
129 K.U. von Raben, R.K. Chang, B.L. Laube: Chem. Phys. Lett. *79*, 465 (1981)
130 J.A. Creighton: "Metal Colloids", in /69/, p. 315
131 C.G. Blatchford, J.R. Campbell, J.A. Creighton: Surf. Sci. *120*, 435 (1982)
132 H. Abe, K. Manzel, W. Schulze, M. Moskovits, D.P. DiLella: J. Chem. Phys. *74*, 792 (1981)
133 M. Moskovits, D.P. DiLella: "Vibrational Spectroscopy of Molecules Adsorbed on Vapor-Deposited Films", in /69/, p. 243
134 T.H. Wood, M.V. Klein: Sol. State Commun. *35*, 263 (1980)
135 M. Moskovits, D.P. DiLella: Chem. Phys. Lett. *73*, 500 (1980)
136 D.P. DiLella, A. Gohin, R.H. Lipson, P. McBreen, M. Moskovits: J. Chem. Phys. *73*, 4282 (1980)
137 P.F. Liao, J.G. Bergmann, D.S. Chemla, A. Wokaun, J. Melngailis, A.M. Hawryluk, N.P. Economou: Chem. Phys. Lett. *82*, 355 (1981)
138 P.F. Liao: "Silver Structures Produced by Microlithography", in /69/, p. 379
139 U. Wenning, B. Pettinger, H. Wetzel: Chem. Phys. Lett. *70*, 49 (1980)
140 G. Laufer, T.F. Schaaf, J.T. Huneke: J. Chem. Phys. *73*, 2973 (1980)
141 W. Schulze, M. Moskovits: unpublished
142 P.A. Lund, R.R. Smarzewski, D.E. Tevault: Chem. Phys. Lett. *89*, 508 (1982)
143 T.H. Wood, M.V. Klein: unpublished
144 T. López-Rios, C. Pettenkofer, I. Pockrand, A. Otto: Surf. Sci. *121*, L 541 (1982)
145 P.F. Liao, B. Stern: Opt. Lett. *7*, 483 (1982)
146 J.J. Kester, T.E. Furtak: Sol. State Commun. *41*, 457 (1982)
147 T.E. Furtak, J.J. Kester: Phys. Rev. Lett. *45*, 1652 (1980)
148 J.C. Tsang, S.S. Jha, J.R. Kirtley: Phys. Rev. Lett. *46*, 1044 (1981)
149 R. Naaman, S.J. Buelow, O. Cheshnovsky, D.R. Herschbach: J. Phys. Chem. *84*, 2692 (1980)
150 L.A. Sanchez, R.L. Birke, J.R. Lombardi: Chem. Phys. Lett. *79*, 219 (1981)

151 B.H. Loo: J. Chem. Phys. *75*, 5955 (1981)
152 B.H. Loo: Sol. State Commun. *43*, 349 (1982)
153 H. Yamada, Y. Yamamoto, N. Tani: Chem. Phys. Lett. *86*, 397 (1982)
154 H. Yamada, N. Tani, Y. Yamamoto: J. Electron Spectrosc. Relat. Phenom. *30*, 13 (1983)
155 J. Heitbaum: Z. Physik. Chemie *105*, 307 (1977)
156 W. Krasser, A.J. Renouprez: Sol. State Commun. *41*, 231 (1982)
157 H. Yamada, Y. Yamamoto: Chem. Phys. Lett. *77*, 520 (1981)
158 W. Krasser, A.J. Renouprez: J. Raman Spectrosc. *8*, 92 (1979)
159 W. Krasser, A. Fadini, A.J. Renouprez: J. Catal. *62*, 94 (1980)
160 W. Krasser, A. Fadini, E. Rozemuller, A.J. Renouprez: J. Mol. Struct. *66*, 135 (1980)
161 W. Krasser, A.J. Renouprez: J. Raman Spectrosc. *11*, 425 (1981)
162 B. Pettinger: private Communication
163 B.H. Loo: J. Electroanal. Chem. *136*, 209 (1982)
164 D.N. Batchelder, N.J. Poole, D. Bloor: Chem. Phys. Lett. *81*, 561 (1981)
165 R. Dornhaus, R.K. Chang: Sol. State Commun. *34*, 811 (1980)
166 K.A. Bunding, M.I. Bell: Surf. Sci. *118*, 329 (1982)
167 E. Koglin, J.M. Sequaris, P. Valenta: J. Mol. Struct. *60*, 421 (1980)
168 D.A. Long: *Raman Spectroscopy* (McGraw-Hill, New York 1977)
169 J. Behringer: "Theories of Resonance Raman Scattering", in Molecular Spectroscopy, ed. by R.F. Barrow, D.A. Long, D.J. Millen, Vol. 2 (The Chemical Society, London 1974) p. 100
170 M.R. Philpott: J. Chem. Phys. *62*, 1812 (1975)
171 S. Efrima, H. Metiu: J. Chem. Phys. *70*, 1602 (1979)
172 M. Moskovits: J. Chem. Phys. *77*, 4408 (1982)
173 P.K. Aravind, E. Hood, H. Metiu: Surf. Sci. *109*, 95 (1981)
174 D.L. Mills, M. Weber: Phys. Rev. *B26*, 1075 (1982)
175 S.L. McCall, P.M. Platzman, P.A. Wolff: Phys. Lett. *77A*, 381 (1980)
176 D.-S. Wang, H. Chew, M. Kerker: Appl. Optics *19*, 2256 (1980)
177 B.J. Messinger, K.U. von Raben, R.K. Chang, P.W. Barber: Phys. Rev. *B24*, 649 (1981)
178 G.S. Agarwal, S.S. Jha, J.C. Tsang: Phys. Rev. *B25*, 2089 (1982)
179 D.-S. Wang, M. Kerker: Phys. Rev. *B24*, 1777 (1981)
180 M. Kerker, D.-S. Wang, H. Chew: Appl. Optics *19*, 3373; 4159 (1980)
181 K. Ohtaka, M. Inoue: J. Phys. C: Sol. State Phys. *15*, 6463 (1982)
182 P.W. Barber, R.K. Chang, H. Massoudi: Phys. Rev. Lett. *50*, 997 (1983)
183 A. Wokaun, J.P. Gordon, P.F. Liao: Phys. Rev. Lett. *48*, 957 (1982)
184 Z. Kotler, A. Nitzan: J. Phys. Chem. *86*, 2011 (1982)
185 M. Kerker, C.G. Blatchford: Phys. Rev. *B26*, 4052 (1982)
186 J.I. Gersten: J. Chem. Phys. *72*, 5779 (1980)
187 J.I. Gersten, A. Nitzen: J. Chem. Phys. *73*, 3023 (1980)
188 F.J. Adrian: Chem. Phys. Lett. *78*, 45 (1981)
189 R. Ruppin: Sol. State Commun. *39*, 903 (1981)
190 M. Moskovits: J. Chem. Phys. *69*, 4159 (1978)
191 M. Moskovits: Sol. State Commun. *32*, 59 (1979)
192 J.C. Maxwell-Garnett: Philos. Trans. (London) *203*, 385 (1904); *205*, 237 (1906)
193 R.H. Doremus: J. Appl. Phys. *37*, 2775 (1966)
194 P.K. Aravind, A. Nitzan, H. Metiu: Surf. Sci. *110*, 189 (1981)
195 B.N.J. Persson, A. Liebsch: "Optical Properties of Inhomogenous Media", unpublished
196 C.Y. Chen, E. Burstein: Phys. Rev. Lett. *45*, 1287 (1980)
197 Z. Kotler, A. Nitzan: Surf. Sci. *130*, 124 (1983)
198 U. Laor, G.C. Schatz: Chem. Phys. Lett. *82*, 566 (1981)
199 U. Laor, G.C. Schatz: J. Chem. Phys. *76*, 2888 (1982)
200 E. Kröger, E. Kretschmann: Z. Physik *237*, 1 (1970)
201 J.M. Elson, R.H. Ritchie: Phys. Rev. *B4*, 4129 (1971)
202 V. Celli, A. Marvin, F. Toigo: Phys. Rev. *B11*, 1779 (1975)
203 A.A. Maradudin, D.L. Mills: Phys. Rev. *B11*, 1392 (1975)
204 P.K. Aravind, H. Metiu: Chem Phys. Lett. *74*, 301 (1980)

205 K. Arya, R. Zeyher, A.A. Maradudin: Sol. State Commun. *42*, 461 (1982)
206 W.H. Weber, G.W. Ford: Phys. Rev. Lett. *44*, 1774 (1980)
207 F.W. King, R.P. Van Duyne, G.C. Schatz: J. Chem. Phys. *69*, 4472 (1978)
208 S. Efrima, H. Metiu: Chem. Phys. Lett. *60*, 59 (1978)
209 S. Efrima, H. Metiu: J. Chem. Phys. *70*, 1939; 2297 (1979)
210 S. Efrima, H. Metiu: Israel J. Chem. *18*, 17 (1979)
211 G.C. Schatz, R.P. Van Duyne: Surf. Sci. *101*, 425 (1980)
212 G.C. Schatz: "The Image Field Effect: How Important is it?", in /69/, p. 35
213 G.L. Eesley, J.R. Smith: Sol. State Commun. *31*, 815 (1979)
214 G.E. Korzeniewski, T. Maniv, H. Metiu: Chem. Phys. Lett. *73*, 212 (1980)
215 G.E. Korzeniewski, T. Maniv, H. Metiu: J. Chem. Phys. *76*, 1564 (1982)
216 P.J. Feibelmann: Phys. Rev. *B22*, 3654 (1980)
217 P.R. Hilton, D.W. Oxtoby: J. Chem. Phys. *72*, 6346 (1980)
218 G.W. Ford, W.H. Weber: Surf. Sci. *109*, 451 (1981)
219 T.K. Lee, J.L. Birman: Phys. Rev. *B22*, 5961 (1980)
220 T.K. Lee, J.L. Birman: J. Raman Spectrosc. *10*, 140 (1981)
221 T.K. Lee, J.L. Birman: "Coupled Excitation Model and Quantum Test of Image Field Effect", in /69/, p. 51
222 T. Maniv, H. Metiu: Surf. Sci. *101*, 399 (1980)
223 T. Maniv, H. Metiu: Chem. Phys. Lett. *79*, 79 (1981)
224 F.R. Aussenegg, M.E. Lippitsch: Chem. Phys. Lett. *59*, 214 (1978)
225 S.L. McCall, P.M. Platzman: Phys. Rev. *B22*, 1660 (1980)
226 A.G. Mal'shukov: Sol. State Commun. *38*, 907 (1981)
227 R.M. Hexter, M.G. Albrecht: Spectrochim. Acta *35A*, 233 (1979)
228 S. Efrima, H. Metiu: Surf. Sci. *92*, 417 (1980)
229 R.R. Chance, A. Prock, R. Silbey: Adv. Chem. Phys. *37*, 1 (1978)
230 G.W. Ford, W.H. Weber: Surf. Sci. *129*, 123 (1983)
231 F.W. King, G.C. Schatz: Chem. Phys. *38*, 245 (1979)
232 E. Burstein, Y.J. Chen, C.Y. Chen, S. Lundquist, E. Tosatti: Sol. State Commun. *29*, 567 (1979)
233 J.I. Gersten, R.L. Birke, J.R. Lombardi: Phys. Rev. Lett. *43*, 147 (1979)
234 H. Ueba: "Induced Resonance Model", in /69/, p. 173
235 K. Arya, R. Zeyher: Phys. Rev. *B24*, 1852 (1981)
236 B.N.J. Persson: Chem. Phys. Lett. *82*, 561 (1981)
237 F.J. Adrian: J. Chem. Phys. *77*, 5302 (1982)
238 H. Ueba: Surf. Sci. *131*, 347 (1983)
239 A. Otto, I. Pockrand, J. Billmann, C. Pettenkofer: "The 'Adatom model': How Important is Atomic Scale Roughness?", in /69/, p. 147
240 R.M. Lazorenko-Manevich, V.V. Marinyuk, Ya.M. Kolotyrkin: Doklady Akad. Nauk. SSSR *244*, 641 (1979)
241 V.V. Marinyuk, R.M. Lazorenko-Manevich, Ya.M. Kolotyrkin: Elektrokhimiya *14*, 1019 (1978)
242 V.V. Marinyuk, R.M. Lazorenko-Manevich: Elektrokhimiya *16*, 332 (1980)
243 V.V. Marinyuk, R.M. Lazorenko-Manevich, Ya.M. Kolotyrkin: Doklady Akad. Nauk. SSSR *253*, 155 (1980)
244 V.V. Marinyuk, R.M. Lazorenko-Manevich, Ya.M Kolotyrkin: J. Electroanal. Chem. *110*, 111 (1980)
245 V.V. Marinyuk, R.M. Lazorenko-Manevich, Ya.M. Kolotyrkin: Elektrokhimiya *17*, 643 (1981)
246 A. Regis, J. Corset: Chem. Phys. Lett. *70*, 305 (1980)
247 G. Blondeau, J. Zerbino, N. Jaffrezic-Renault: J. Electroanal. Chem. *112*, 127 (1980)
248 M.W. Howard, R.P. Cooney, A.J. McQuillan: J. Raman Spectrosc. *9*, 273 (1980)
249 R.P. Cooney, M.W. Howard, M.R. Mahoney, R.P. Mernagh: Chem. Phys. Lett. *79*, 459 (1981)
250 M.W. Howard, R.P. Cooney: Chem. Phys. Lett. *87*, 299 (1982)
251 P.K.K. Pandey, G.C. Schatz: Chem. Phys. Lett. *88*, 193 (1982)
252 P.K.K. Pandey, G.C. Schatz: Chem. Phys. Lett. *91*, 286 (1982)
253 P.K.K. Pandey, G.C. Schatz: J. Electron Spectrosc. Relat. Phenom. *29*, 351 (1983)
254 T.L. Ferrell: Phys. Rev. *B25*, 2930 (1982)

255 C.A. Murray, D.L. Allara, M. Rhinewine: Phys. Rev. Lett. *46*, 57 (1981)
256 C.A. Murray, D.L. Allara: J. Chem. Phys. *76*, 1290 (1982)
257 C,A, Murray, D.L. Allara, A.F. Hebard, F.J. Padden, Jr.: Surf. Sci. *119*, 449 (1982)
258 C.A. Murray: "Molecule-Silver Separation Dependence", in /69/, p. 203
259 H. Seki: J. Vac. Sci. Technol. *18*, 633 (1981)
260 G.L. Eesley, J.M. Burkstrand: Phys. Rev. *B24*, 582 (1981)
261 J.A. Creighton: Surf. Sci. *124*, 209 (1983)
262 J.E. Demuth, P.N. Sanda: Phys. Rev. Lett. *47*, 57 (1981)
263 D. Schmeisser, J.E. Demuth, Ph. Avouris: Chem. Phys. Lett. *87*, 324 (1982)
264 J. Billmann, A. Otto: Sol. State Commun. *44*, 105 (1982)
265 A. Otto: "On the 'Chemical Effect' in Surface Enhanced Raman Scattering", in Proc. VIIIth Int. Conf. on Raman Spectroscopy, Bordeaux, 1982; ed. by J. Lascombe, Pham. V. Huong (Wiley, Chichester 1982) p. 49
266 H. Seki, M.R. Philpott: J. Chem. Phys. *73*, 5376 (1980)
267 I. Pockrand: Chem. Phys. Lett. *92*, 509 (1982)
268 P.N. Sanda, J.E. Demuth, J.C. Tsang, J.M. Warlaumont: "Coverage Dependence", in /69/, p. 189
269 G.L. Eesley, D.L. Simon: J. Vac. Sci. Technol. *18*, 629 (1981)
270 C. Pettenkofer: unpublished
271 T.H. Wood: Phys. Rev. *B24*, 2289 (1981)
272 I. Pockrand: unpublished
273 Ü. Ertürk, I. Pockrand, A. Otto: Surf. Sci. *131*, 367 (1983)
274 B.C. Allen: Trans. Metall. Soc. AIME *227*, 1175 (1983)
275 J. Friedel: Ann. Phys. (Paris) *1*, 257 (1976)
276 T. López-Rios: private communication
277 J. Eickmanns, A. Goldmann, A. Otto: Surf. Sci. *127*, 163 (1983)
278 P.G. Hall, D.A. King: Surf. Sci. *36*, 810 (1973)
279 J. Küppers, K. Wandelt, G. Ertl: Phys. Rev. Lett. *43*, 928 (1979)
280 J. Hulse, J. Küppers, K. Wandelt, G. Ertl: Applic. Surf. Sci. *6*, 453 (1980)
281 I. Pockrand: Chem. Phys. Lett. *92*, 514 (1982)
282 T. López-Rios, Y. Borensztein, G. Vuye: J. Physique *44*, L 99 (1983)
283 I. Pockrand: unpublished
284 I. Pockrand: Surf. Sci. *72*, 577 (1978)
285 O. Hunderi, H.P. Myers: J. Phys. F: Metal Phys. *3*, 683 (1973)
286 O. Hunderi: J. Physique *38*, C5 - 89 (1977)
287 O. Hunderi: Surf. Sci. *96*, 1 (1980)
288 J.P. Chauvineau: unpublished
289 D. Schumacher, D. Stark: Surf. Sci. *123*, 384 (1982)
290 L.J. Cuddy, E.S. Machlin: Phil. Mag. *7*, 745 (1962)
291 M. Doyama, J.S. Koehler: Phys. Rev. *127*, 21 (1962)
292 L.M. Clarebrough, R.C. Segall, M.H. Loretto, E.M. Hargreaves: Phil. Mag. *9*, 377 (1964)
293 H. Mehrer, A. Seeger: Phys. Stat. Sol. *39*, 647 (1970)
294 C.Y. Chen, I. Davoli, G. Ritchie, E. Burstein: Surf. Sci. *101*, 363 (1980)
295 R.S. Bennett, G.D. Scott: J. Opt. Soc. Am. *40*, 203 (1950)
296 A.M. Glass, P.F. Liao, J.G. Bergman, D.H. Olson: Opt. Lett. *5*, 368 (1980)
297 S. Yoshida, T. Yamaguchi, A. Kinbara: J. Opt. Soc. Am. *61*, 62 (1971)
298 S. Yoshida, T. Yamaguchi, A. Kinbara: J. Opt. Soc. Am. *61*, 463 (1971)
299 T. Yamaguchi, S. Yoshida, A. Kinbara: Thin Sol. Films *21*, 173 (1974)
300 T. Yamaguchi, S. Yoshida, A. Kinbara: J. Opt. Soc. Am. *64*, 1563 (1974)
301 I. Pockrand: Phys. Lett. *49A*, 259 (1974)
302 E. Rosengart, I. Pockrand: Opt. Lett. *1*, 194 (1977)
303 I. Pockrand: "Optische Eigenschwingungen in dünnen Schichten mit periodisch gestörter Oberfläche"; Dissertation, Universität Hamburg (1976)
304 H. Raether: "Surface Plasmons and Roughness", in *Surface Polaritons*, ed. by V.M. Agranovich, D.L. Mills (North - Holland, Amsterdam 1982) p. 331
305 R. Petit (ed.): *Electromagnetic Theory of Gratings*, Topics in Current Physics, Vol. 22 (Springer, Berlin, Heidelberg, New York 1980)
306 M.C. Hutley: *Diffraction Gratings* (Academic Press, London 1982)

307 D.A. Zwemer, C.V. Shank, J.E. Rowe: Chem. Phys. Lett. *73*, 201 (1980)
308 H. Abe, W. Schulze, B. Tesche: Chem. Phys. *47*, 95 (1980)
309 S. Yatsuya, S. Kasukabe, R. Uyeda: Jap. J. Appl. Phys. *12*, 1675 (1973)
310 M. Rappaz, F. Faes: J. Appl. Phys. *46*, 3273 (1975)
311 W. Schulze, H.U. Becker, H. Abe: Chem. Phys. *35*, 177 (1978)
312 T. Welker, T.P. Martin: J. Chem. Phys. *70*, 5683 (1979)
313 E.P. Parry: J. Catal. *2*, 371 (1963)
314 P.J. Hendra, J.R. Horder, E.J. Loader: Chem. Commun. 563 (1970)
315 P.J. Hendra, I.D.M. Turner, E.J. Loader, M. Stacey: J. Phys. Chem. *78*, 300 (1974)
316 E.K. Rideal: *Concepts in Catalysis* (Academic Press, London 1968)
317 H.A. Benesi, B.H.C. Winquist: Adv. Catal. *27*, 97 (1978)
318 M. Boudart: "Concepts of Heterogeneous Catalysis", in *Interactions on Metal Surfaces*, ed. by R. Gomer, Topics in Applied Physics, Vol. 4 (Springer, Berlin, Heidelberg, New York 1975) p. 275
319 T.E. Madey, J.T. Yates Jr., D.R. Sandstrom, F.J.H. Voorhoeve: "Catalysis by Solid Surfaces", in Treatise on Solid State Chemistry, ed. by N.B. Hannay, Vol. 68 (Plenum, New York 1976)
320 G.A. Somorjai: Adv. Catal. *29*, 1 (1980)
321 G.A. Somorjai: Surf. Sci. *89*, 496 (1979)
322 B. Bak, L. Hansen, J. Rastrup-Andersen: J. Chem. Phys. *22*, 2013 (1954)
323 B. Bak, L. Hansen-Nygard, J. Rastrup: J. Mol. Spectrosc. *2*, 361 (1958)
324 F. Mata, M.J. Quintana, G.O. Sørensen: J. Mol. Struct. *42*, 1 (1977)
325 J.L. Gland, G.A. Somorjai: Surf. Sci. *38*, 157 (1973)
326 B.B. DeMore, W.S. Wilcox, J.H. Goldstein: J. Chem. Phys. *22*, 876 (1954)
327 H.F. Hameka, A.M. Liquori: Mol. Phys. *1*, 9 (1958)
328 D.A. Long, F.S. Murfin, J.L. Hales, W. Kynaston: Trans. Faraday Soc. *53*, 1171 (1957)
329 D.P. DiLella, H.P. Stidham: J. Raman Spectrosc. *9*, 90 (1980)
330 L. Corrsin, B.J. Fax, R.C. Lord: J. Chem. Phys. *21*, 1170 (1953)
331 J.K. Wilmshurst, H.J. Bernstein: Can. J. Chem. *35*, 1183 (1957)
332 *Raman / IR Atlas of Organic Compounds* (Verlag Chemie, Weinheim 1974)
333 H.D. Stidham, D.P. DiLella: J. Raman Spectrosc. *8*, 180 (1979)
334 H. Takahashi, K. Mamola, E.K. Plyler: J. Mol. Spectrosc. *21*, 217 (1966)
335 J. Loisel, V. Lorenzelli: J. Mol. Struct. *1*, 157 (1967)
336 E. Castelluci, G. Sbrana, F.D. Verderame: J. Chem. Phys. *51*, 3762 (1969)
337 D.A. Long, F.S. Marfin, E.L. Thomas: Trans. Faraday Soc. *59*, 12 (1963)
338 D.A. Long, E.L. Thomas: Trans. Faraday Soc. *59*, 783 (1963)
339 S. Suzuki, W.J. Orville-Thomas: J. Mol. Struct. *37*, 321 (1977)
340 L.P. Bicelli: Nuovo Cimento *9*, 184 (1958)
341 L.P. Bicelli: Instituto Lombardi (Rend. Sc.) *A92*, 536 (1958)
342 M. Goldstein, E.F. Mooney, A. Anderson, H.A. Gebbie: Spectrochim. Acta *21*, 105 (1965)
343 R.J.H. Clark, C.S. Williams: Inorg. Chem. *4*, 350 (1965)
344 C.W. Frank, L.B. Rogers: Inorg. Chem. *5*, 615 (1966)
345 M. Goldstein, W.D. Unsworth: Inorg. Chim. Acta *4*, 342 (1970)
346 Y. Saito, M. Cordes, K. Nakamoto: Spectrochim. Acta *A28*, 1459 (1973)
347 S. Akyüz, A.B. Dempster, R.L. Morehouse, S. Suzuki: J. Mol. Struct. *17*, 105 (1973)
348 K.K. Innes, J.P. Byrne, I.G. Ross: J. Mol. Spectrosc. *22*, 125 (1967)
349 J.P. Doering, J.H. Moore Jr.: J. Chem. Phys. *56*, 2176 (1972)
350 Ph. Avouris, J.E. Demuth: J. Chem. Phys. *75*, 4783 (1981)
351 J.E. Demuth, Ph. Avouris, P.N. Sanda: J. Vac. Sci. Technol. *20*, 588 (1982)
352 R.S. Mulliken, W.B. Person: *Molecular Complexes* (Wiley, New York 1969)
353 J.E. Demuth, Ph. Avouris, D. Schmeisser: J. Electron. Spectrosc. Relat. Phenom. *29*, 163 (1983)
354 Ph. Avouris, J.E. Demuth: J. Chem. Phys. *75*, 5953 (1981)
355 S.R. Kelemen, A. Kaldor: Chem. Phys. Lett. *73*, 205 (1980)
356 B.J. Bandy, D.R. Lloyd, N.V. Richardson: Surf. Sci. *89*, 344 (1979)
357 F.P. Netzer, E. Bertel, J.A.D. Matthew: Surf. Sci. *92*, 43 (1980)

358 I. Pockrand: unpublished
359 D.E. Tevault, R.R. Smardzewski: "Infrared Spectra of Silver Atom/Cluster Complexes with Pyridine and Their Relationship to the Surface Enhanced Raman Effect", to be published
360 H. Seki: J. Electroanal. Chem. *150*, 425 (1983)
361 J.I. Gersten, A. Nitzan: "Electromagnetic Theory: A Spheroidal Model", in /69/, p. 89
362 M. Kerker, D.-S. Wang, H. Chew, O. Siiman, L.A. Bumm: "Enhanced Raman Scattering by Molecules Adsorbed at the Surface of Colloidal Particles", in /69/, p. 109
363 P.B. Johnson, R.W. Christy: Phys. Rev. *B12*, 4370 (1972)
364 P. Rouard, A. Meessen: "Optical Properties of Thin Metal Films", in Progress in Optics, ed. by E. Wolff, Vol. XV (North-Holland, Amsterdam 1977) p. 79
365 R.R. Bilboul: J. Phys. D: Appl. Phys. *2*, 921 (1969)
366 T.H. Wood, D.A. Zwemer, C.V. Shank, J.E. Rowe: Chem. Phys. Lett. *82*, 5 (1981)
367 K. Kishi, K. Chinomi, Y. Inoue, S. Ikeda: J. Catal. *60*, 228 (1979)
368 D.M. Kolb, W. Boeck, K.M. Ho, S.H. Liu: Phys. Rev. Lett. *47*, 1921 (1982)
369 H.D. Ladouceur, D.E. Tevault, R.R. Smardzewski: J. Chem. Phys. *78*, 980 (1983)
370 H. Seki: Sol. State Commun. *42*, 695 (1982)
371 B. Pettinger: private communication
372 A. Campion, D.R. Mullins: Chem. Phys. Lett. *94*, 576 (1983)
373 H. Seki: J. Electron Spectrosc. Relat. Phenom. *29*, 413 (1983)
374 I. Pockrand: unpublished
375 W. Schlemminger, D. Stark: unpublished
376 J.P. Chauvineau: Surf. Sci. *93*, 471 (1980)
377 R. Smoluchowski: Phys. Rev. *60*, 661 (1941)
378 L.L. Kesmodel, L.M. Falicov: Sol. State Commun. *16*, 1201 (1975)
379 Y.W. Tsang, L.M. Falicov: J. Phys. C: Sol. State Phys. *9*, 51 (1976)
380 B. Krahl-Urban, E.A. Niekisch, H. Wagner: Surf. Sci. *64*, 52 (1977)
381 K. Besocke, B. Krahl-Urban, H. Wagner: Surf. Sci. *68*, 39 (1977)
382 H. Ibach: Surf. Sci. *53*, 444 (1975)
383 H. Hopster, H. Ibach, G. Comsa: J. Catal. *46*, 37 (1977)
384 H. Wagner: "Physical and Chemical Properties of Stepped Surfaces", in *Solid Surface Physics*, Springer Tracts in Modern Physics, ed. by G. Höhler, Vol. 85 (Springer, Berlin, Heidelberg, New York 1979) p. 151
385 C. Backx, C.P.M. de Groot, P. Biloen: Applic. Surf. Sci. *6*, 256 (1980)
386 A. Otto: private communication
387 C.N. Scatterfield: *Heterogeneous Catalysis in Practice* (McGraw-Hill, New York 1980)
388 J.R. Anderson: *Structure of Metallic Catalysts* (Academic Press, London 1975)
389 C.L. Thomas: *Catalytic Processes and Proven Catalysts* (Academic Press, New York 1970)
390 X.E. Verykos, F.P. Stein, R.W. Coughlin: Catal. Rev. Sci. Eng. *22*, 197 (1980)
391 W.M.H. Sachtler, C. Backx, R.A. van Santen: Catal. Rev. Sci. Eng. *23*, 127 (1981)
392 J.V. Porcelli: Catal. Rev. Sci. Eng. *23*, 151 (1981)
393 J.C. Zomerdijk, M.W. Hall: Catal. Rev. Sci. Eng. *23*, 163 (1981)
394 C.N. Stewart, G. Ehrlich: J. Chem. Phys. *62*, 4672 (1975)
395 J.T. Yates Jr., T.E. Madey: Surf. Sci. *28*, 437 (1971)
396 R. Suhrmann, H.J. Busse, G. Wedler: Z. Phys. Chemie (Leipzig) *229*, 10 (1965)
397 T.E. Madey, J.T. Yates Jr.: Surf. Sci. *76*, 397 (1978)
398 K. Horn, J. Pritchard: Surf. Sci. *52*, 437 (1975)
399 K. Horn: "IR Reflection-Absorption Spectroscopy of Mono- and Multilayers of Ethane on Pt(111)", in Proc. Int. Conf. Vibrations in Adsorbed Layers, Jülich, June 1978, ed. by H. Ibach, S. Lehwald; Jül.-Conf. 26, p. 140
400 I. Pockrand: Vhdlg. der DPG *5*, 944 (1982)
401 D.P. DiLella, M. Moskovits: J. Phys. Chem. *85*, 2042 (1981)
402 K. Manzel, W. Schulze, M. Moskovits: Chem. Phys. Lett. *85*, 183 (1982)
403 K. Kochitsu: J. Chem. Phys. *44*, 906 (1966)

404 J.L. Duncan: Molec. Phys. *28*, 1177 (1974)
405 G.K.T. Conn, G.B.B.M. Sutherland: Proc. Roy. Soc. *A172*, 172 (1939)
406 W.S. Gallaway, E.F. Barker: J. Chem. Phys. *10*, 88 (1942)
407 R.L. Arnett, B.L. Crawford Jr.: J. Chem. Phys. *18*, 118 (1950)
408 J.L. Duncan, E. Hamilton: J. Mol. Struct. *76*, 65 (1981)
409 M. de Hemptinne, J. Jungers, J.M. Delfosse: J. Chem. Phys. *6*, 319 (1938)
410 C. Brecher, R.S. Halford: J. Chem. Phys. *35*, 1109 (1961)
411 M.E. Jacox: J. Chem. Phys. *36*, 140 (1962)
412 G.R. Elliot, G.E. Leroi: J. Chem. Phys. *59*, 1217 (1973)
413 J.L. Duncan, D.C. McKean, P.D. Mallinson: J. Mol. Spectrosc. *45*, 221 (1973)
414 E. Rytter, D.M. Gruen: Spectrochim. Acta *35A*, 199 (1979)
415 M.J. Grogan, K. Nakamoto: J. Amer. Chem. Soc. *88*, 5454 (1966)
416 J. Pradilla-Sorzano, J.P. Fackler Jr.: J. Mol. Spectrosc. *22*, 80 (1967)
417 J. Hiraishi: Spectrochim. Acta *25A*, 749 (1969)
418 D.B. Powell, J.G.V. Scott, N. Sheppard: Spectrochim. Acta *28A*, 327 (1972)
419 H. Huber, D. McIntosh, G.A. Ozin: J. Organometal. Chem. *112*, C50 (1976)
420 D. McIntosh, G.A. Ozin: J. Organometal. Chem. *121*, 127 (1976)
421 G.A. Ozin, H. Huber, D. McIntosh: Inorg. Chem. *16*, 3070 (1977)
422 D.F. McIntosh, G.A. Ozin, R.P. Messmer: Inorg. Chem. *19*, 3321 (1980)
423 P.H. Kasai, D. McLeod, T. Watanabe: J. Amer. Chem. Soc. *102*, 179 (1980)
424 G. Herzberg: *Molecular Spectra and Molecular Structure. III. Electronic Spectra and Electronic Structure of Polyatomic Molecules* (van Nostrand, New York 1966)
425 K.K. Innes: "Electronic Spectra", in *Molecular Spectroscopy: Modern Research*, ed. by K. Narahari Rao, C.W. Mathews (Academic Press, New York 1972) p. 179
426 A.J. Merer, R.S. Mullikan: Chem. Rev. *69*, 639 (1969)
427 W.C. Price: Phys. Rev. *47*, 444 (1935)
428 E.N. Lassettre, A. Skerbele, M.A. Dillon, K.J. Ross: J. Chem. Phys. *48*, 5066 (1968)
429 J.P. Doering, A.J. Williams III: J. Chem. Phys. *47*, 4180 (1967)
430 H. Ibach, S. Lehwald: J. Vac. Sci. Technol. *15*, 407 (1978)
431 J.E. Demuth: Surf. Sci. *84*, 315 (1979)
432 N.D.S. Canning, M.D. Baker, M.A. Chesters: Surf. Sci. *111*, 441 (1981)
433 T.E. Felter, W.H. Weinberg: Surf. Sci. *103*, 265 (1981)
434 L.H. Dubois, D.G. Castner, G.A. Somorjai: J. Chem. Phys. *72*, 5234 (1980)
435 W. Erley, A.M. Baro, H. Ibach: Surf. Sci. *120*, 273 (1982)
436 S. Lehwald, H. Ibach: Surf. Sci. *89*, 425 (1979)
437 J.E. Demuth, D.E. Eastman: Phys. Rev. Lett. *32*, 1123 (1974)
438 J.A. Gates, L.L. Kesmodel: Surf. Sci. *120*, L 461 (1982)
439 L.L. Kesmodel, J.A. Gates: Surf. Sci. *111*, L 747 (1981)
440 J.A. Gates, L.L. Kesmodel: Surf. Sci. *124*, 68 (1983)
441 M.A. van Howe, R.J. Koestner, G.A. Somorjai: J. Vac. Sci. Technol. *20*, 886 (1982)
442 L.L. Kesmodel, L.H. Dubois, G.A. Somorjai: Chem. Phys. Lett. *56*, 267 (1978)
443 J.E. Demuth: Surf. Sci. *80*, 367 (1979)
444 J.E. Demuth, H. Ibach, S. Lehwald: Phys. Rev. Lett. *40*, 1044 (1978)
445 J.C. Bertolini, J. Rousseau: Surf. Sci. *83*, 531 (1979)
446 M. Ito, Y. Mori, T. Kato, W. Suëtaka: Applic. Surf. Sci. *2*, 543 (1979)
447 R. Ducros, M. Housley, M. Alnot, A. Cassuto: Surf. Sci. *71*, 433 (1978)
448 R. Ducros, M. Housley, G. Piquard, M. Alnot: Surf. Sci. *109*, 235 (1981)
449 C. Backx, C.P.M. de Groot: Surf. Sci. *115*, 382 (1982)
450 C. Nyberg, C.G. Tengstål, S. Andersson: Chem Phys. Lett. *87*, 87 (1982)
451 M. Ito, W. Suëtaka: Surf. Sci. *62*, 308 (1977)
452 J.E. Demuth: IBM J. Res. Develop. *22*, 265 (1978)
453 T.E. Felter, W.H. Weinberg, P.A. Zhdan, G.K. Boreskov: Surf. Sci. *97*, L 313 (1980)
454 S.R. Kelemen, T.E. Fischer: Surf. Sci. *102*, 45 (1981)
455 G. Rovida, F. Pratesi, E. Ferroni: Applic. Surf. Sci. *5*, 121 (1980)
456 M.A. Barteau, R.J. Madix: Surf. Sci. *103*, L 171 (1981)
457 I.E. Wachs, S.R. Kelemen: "Mechanism of the Interaction of Ethylene with Atomic Oxygen on a Silver Surface", in Proc. 7th Intern. Congress on Catalysis, Tokyo 1980, p. 682

458 T.H. Wood, D.A. Zwemer: J. Vac. Sci. Technol. *18*, 649 (1981)
459 I. Pockrand: Surf. Sci. *126*, 192 (1983)
460 I. Pockrand: to be published
461 A.V. Bobrov, A.N. Gass, O.I. Kapusta, N.M. Omel'yanovskaya: JETP Lett. *35*, 626 (1982)
462 O. Ertürk, I. Pockrand, A. Otto: to be published
463 B.A. Sexton: Appl. Phys. *A26*, 1 (1981)
464 B.N.J. Persson, R. Ryberg: Phys. Rev. *B24*, 6954 (1981)
465 U. Fano: Phys. Rev. *124*, 1866 (1961)
466 R.J. Madix: Applic. Surf. Sci. *14*, 41 (1982)
467 L.M. Sverdlov, M.A. Kovner, E.P. Krainov: *Vibrational Spectra of Polyatomic Molecules* (Wiley, New York 1974)
468 R.S. Rasmussen, R.R. Brattain: J. Chem. Phys. *15*, 120 (1947)
469 R.C. Lord, P. Venkateswarlu: J. Opt. Soc. Am. *43*, 1079 (1953)
470 W. Lüttke, S. Braun: Ber. Bunsenges. Phys. Chem. *71*, 34 (1967)
471 C.M. Pathak, W.H. Fletcher: J. Mol. Spectrosc. *31*, 32 (1969)
472 I.W. Levin, R.A.R. Pearce: J. Mol. Spectrosc. *49*, 91 (1974)
473 I.W. Levin, R.A.R. Pearce, W.C. Harris: J. Chem. Phys. *59*, 3048 (1973)
474 K. Nakamoto: *Infrared and Raman Spectra of Inorganic and Coordination Compounds* (Wiley-Interscience, New York 1978)
475 E. Maslowsky Jr.: *Vibrational Spectra of Organometallic Compounds* (Wiley-Interscience, New York 1977)
476 M.A. Bennett: Chem. Rev. *62*, 611 (1962)
477 R.G. Guy, B.L. Shaw: Adv. Inorg. Chem. Radiochem. *4*, 77 (1962)
478 C.D.M. Beverwijk, G.J.M. van der Kerk, A.J. Leusink, J.G. Noltes: Organometal. Chem. Rev. A, *5*, 215 (1970)
479 H.W. Quinn, J.S. McIntyre, D.J. Peterson: Can. J. Chem. *43*, 2896 (1965)
480 H.J. Taufen, M.J. Murray, F.F. Cleveland: J. Amer. Chem. Soc. *63*, 3500 (1941)
481 J.A.R. Samson, F.F. Marmo, K. Watanabe: J. Chem. Phys. *36*, 783 (1962)
482 J.T. Gary, L.W. Pickett: J. Chem. Phys. *22*, 599 (1954)
483 B.A. Morrow, N. Sheppard: Proc. Roy. Soc. *A311*, 415 (1969)
484 J. Wojtczak, R. Queau, R. Poilblanc: J. Catal. *37*, 391 (1975)
485 J.L. Gland, G.A. Somorjai: Adv. Colloid Interface Sci. *5*, 205 (1976)
486 M.A. Van Hove, L.H. Dubois, R.J. Koestner, G.A. Somorjai: "The Structure of CO, CO_2 and Small Hydrocarbon Molecules (Acetylene, Ethylene, Propylene, Methylacetylene) Adsorbed on Rh(111) and Pt(111) Studied by LEED and HREELS", in Proc. 4th Int. Conf. on Solid Surfaces, Cannes, 1980; p. 287
487 H.H. Voge, C.R. Adams: Adv. Catal. *17*, 151 (1967)
488 R.K. Grasselli, J.D. Burrington: Adv. Catal. *30*, 133 (1981)
489 T.L. Brown: Chem. Rev. *58*, 581 (1958)
490 E.E. Bell, H.H. Nielson: J. Chem. Phys. *18*, 1382 (1950)
491 T.A. Wiggins, E.K. Plyler, E.D. Tidwell: J. Opt. Soc. Am. *51*, 1219 (1961)
492 E.D. Tidwell, E.K. Plyler: J. Opt. Soc. Am. *52*, 656 (1962)
493 G.L. Bottger, D.F. Eggers: J. Chem. Phys. *40*, 2010 (1964)
494 R. Mason, K.M. Thomas: Ann. NY Acad. Sci. *238*, 225 (1974)
495 D. Blake, G. Calvin, G.E. Coates: Proc. Chem. Soc. 396 (1959)
496 A.E. Comyns, H.J. Lucas: J. Amer. Chem. Soc. *79*, 4341 (1957)
497 R. Vestin, A. Somersalo, B. Mueller: Acta Chem. Scand. *7*, 745 (1953)
498 G.E. Coates, C. Parkin: J. Inorg. Nucl. Chem. *22*, 59 (1961)
499 R. Nast, H. Schindel: Z. anorg. allg. Chemie *326*, 201 (1963)
500 G.A. Ozin, D.F. McIntosh, W.J. Power, R.P. Messmer: Inorg. Chem. *20*, 1782 (1981)
501 C.K. Ingold, G.W. King: J. Chem. Soc. 2702 (1953)
502 J.T. Hougen, J.K.G. Watson: Can. J. Phys. *43*, 298 (1965)
503 S. Trajmar, J.K. Rice, P.S.P. Wei, A. Kuppermann: Chem. Phys. Lett. *1*, 703 (1968)
504 J.-C. Bertolini, J. Massardier, G. Dalmai-Imelik: J.C.S. Faraday I *74*, 1720 (1978)
505 J.E. Demuth, H. Ibach: Surf. Sci. *85*, 365 (1979)
506 T.E. Fischer, S.R. Kelemen, H.P. Bonzel: Surf. Sci. *64*, 157 (1977)

507 L.L. Kesmodel, P.C. Stair, R.C. Baetzold, G.A. Somorjai: Phys. Rev. Lett. *36*, 1316 (1976)
508 H. Ibach, H. Hopster, B. Sexton: Appl. Phys. *14*, 21 (1977)
509 J.A. Gates, L.L. Kesmodel: J. Chem. Phys. *76*, 4281 (1982)
510 G. Brodén, T. Rhodin, W. Capehart: Surf. Sci. *61*, 143 (1976)
511 T.E. Fischer, S.R. Kelemen: Surf. Sci. *74*, 47 (1978)
512 C. Backx, R.F. Willis, B. Feuerbacher, B. Fitton: Surf. Sci. *68*, 516 (1977)
513 J.E. Demuth: Surf. Sci. *93*, 127 (1980)
514 J.E. Demuth, H. Ibach: Surf. Sci. *78*, L 238 (1978)
515 S. Lehwald, W. Erley, H. Ibach, H. Wagner: Chem Phys. Lett. *62*, 360 (1979)
516 N. Sheppard, J.W. Ward: J. Catal. *15*, 50 (1969)
517 K.Y. Yu, W.E. Spicer, I. Lindau, P. Pianetta, S.F. Lin: Surf. Sci. *57*, 157 (1976)
518 C.P. Nash, R.P. De Sieno: J. Phys. Chem. *69*, 2139 (1965)
519 L.H. Little, N. Sheppard, D.J.C. Yates: Proc. Roy. Soc. *A259*, 242 (1961)
520 I.E. Wachs, S.R. Kelemen: J. Catal. *68*, 213 (1981)
521 M.A. Barteau, R.J. Madix: Surf. Sci. *115*, 355 (1982)
522 E.M. Stuve, R.J. Madix, B.A. Sexton: Surf. Sci. *123*, 491 (1982)
523 I. Pockrand, C. Pettenkofer, A. Otto: J. Electron Spectrosc. Relat. Phenom. *29*, 409 (1983)
524 W.F. Colby: Phys. Rev. *47*, 388 (1935)
525 I. Pockrand: to be published
526 J. Goubeau, O. Beurer: Z. anorg. allg. Chem. *310*, 110 (1961)
527 V.T. Aleksanyan, I.A. Garbuzova, I.R. Gol'ding, A.M. Sladkov: Spectrochim. Acta *31A*, 517 (1975)
528 L.L. Kesmodel, R.C. Baetzold, G.A. Somorjai: Surf. Sci. *66*, 299 (1977)
529 P.C. Painter, J.L. Koenig: Spectrochim. Acta *33A*, 1003 (1977)
530 P.C. Painter, J.L. Koenig: Spectrochim. Acta *33A*, 1019 (1977)
531 E.B. Wilson Jr.: Phys. Rev. *45*, 706 (1934)
532 H.P. Fritz, W. Lüttke, H. Stammreich, R. Forneris: Spectrochim. Acta *17*, 1068 (1961)
533 H.P. Fritz, E.O. Fischer: J. Organometal. Chem. *7*, 121 (1967)
534 L. Schäfer, J.F. Southern, S.J. Cyvin: Spectrochim. Acta *27A*, 1083 (1970)
535 S.J. Cyvin, J. Brunvoll, L. Schäfer: J. Chem. Phys. *54*, 1517 (1971)
536 E.O. Fischer, W. Hafner: Z. Naturforschg. *10b*, 665 (1955)
537 H.G. Smith, R.E. Rundle: J. Amer. Chem. Soc. *80*, 5075 (1958)
538 P.C. Stair, G.A. Somorjai: J. Chem. Phys. *67*, 4361 (1977)
539 P. Hofmann, K. Horn, A.M. Bradshaw: Surf. Sci. *105*, L 260 (1981)
540 N.V. Richardson, N.R. Palmer: Surf. Sci. *114*, L 1 (1982)
541 H. Jobic, J. Tomkinson: Surf. Sci. *95*, 496 (1980)
542 D.M. Haaland: Surf. Sci. *102*, 405 (1981)
543 D.M. Haaland: Surf. Sci. *111*, 555 (1981)
544 J.C. Bertolini, G. Dalmai-Imelik, J. Rousseau: Surf. Sci. *67*, 478 (1977)
545 J.C. Bertolini, J. Rousseau: Surf. Sci. *89*, 467 (1979)
546 S. Lehwald, H. Ibach, J.E. Demuth: Surf. Sci. *78*, 577 (1978)
547 J.E. Demuth, P.N. Sanda, J.M. Warlaumont, J.C. Tsang, K. Christmann: "High Resolution Electron Energy Loss and Surface Enhanced Raman Studies of Pyridine and Benzene on Ag(111)" in *Vibrations at Surfaces*, ed. by R. Caudano, J.-M. Gilles, A.A. Lucas (Plenum, New York 1982) p. 391
548 D.A. Weitz, S. Garoff, J.I. Gersten, A. Nitzan: J. Chem. Phys. *78*, 5324 (1983)
549 D.A. Weitz, S. Garoff, J.I. Gersten, A. Nitzan: J. Electron Spectrosc. Relat. Phenom. *29*, 363 (1983)
550 J.C. Tsang, Ph. Avouris, J.R. Kirtley: Chem. Phys. Lett. *94*, 172 (1983)
551 J.C. Tsang, Ph. Avouris, J.R. Kirtley: J. Chem. Phys. *79*, 493 (1983)
552 R. Dornhaus: "Surface Enhanced Raman Scattering from Different Types of Metal Substrates", in *Vibrations at Surfaces*, ed. by R. Caudano, J.-M. Gilles, A.A. Lucas (Plenum, New York 1982) p. 445
553 D.A. Zwemer, J.E. Rowe, C.V. Shank, C.A. Murray: "Enhanced Raman Scattering in Ultrahigh Vacuum" in Proc. VIIth Int. Conf. on Raman Spectroscopy, Ottawa, 1980; ed. by W.E. Murphy (North-Holland, Amsterdam 1980) p. 414

554 M.A. Vannice: J. Catal. *37*, 449 (1975)
555 M.A. Vannice: J. Catal. *37*, 462 (1975)
556 G.A. Somorjai: Catal. Rev. Sci. Eng. *23*, 189 (1981)
557 H.H. Storck, N. Golumbic, R.B. Anderson: in *The Fischer–Tropsch and Related Reactions* (Wiley, New York 1951)
558 P. Biloen, W.M.H. Sachtler: Adv. Catal. *30*, 165 (1981)
559 F.G. Dwyer: Catal. Rev. *6*, 261 (1972)
560 N. Sheppard, T.T. Nguyen: "The Vibrational Spectra of Carbon Monoxide Chemisorbed on the Surfaces of Metal Catalysts - A Suggested Scheme of Interpretation", in *Advances in Infrared and Raman Spectroscopy*, Vol. 5, ed. by R.J.H. Clark, R.E. Hester (Heyden, London 1978) p. 67
561 A.M. Bradshaw: Surf. Sci. *80*, 125 (1979)
562 G. Herzberg: *Molecular Spectra and Molecular Structure. I. Spectra of Diatomic Molecules* (van Nostrand, Princeton 1950)
563 S. Green: Adv. Chem. Phys. *25*, 179 (1974)
564 P.S. Braterman: *Metal Carbonyl Spectra* (Academic Press, London 1975)
565 H. Huber, E.P. Kündig, M. Moskovits, G.A. Ozin: J. Amer. Chem. Soc. *97*, 2097 (1975)
566 Y. Souma, J. Iyodo, H. Sano: Inorg. Chem. *15*, 968 (1976)
567 D. McIntosh, G.A. Ozin: J. Amer. Chem. Soc. *98*, 3167 (1976)
568 A. Neppel, J.P. Hickey, I.S. Butler: J. Raman Spectrosc. *8*, 57 (1979)
569 P. Thiry: "Vibrations measured at Metal Surfaces by EELS: A Review Table", in *Vibrations at Surfaces*, ed. by R. Caudano, J.-M. Gilles, A.A. Lucas (Plenum Press, New York 1982) p. 231
570 J. Darville: "Infrared Spectra of Species Adsorbed on Films or Well-Defined Surfaces: A Tabulation of the Investigated Systems", in *Vibrations at Surfaces*, ed. by R. Caudano, J.-M. Gilles, A.A. Lucas (Plenum Press, New York 1982) p. 341
571 K. Horn, C. Mariani: Vhdlg. der DPG *5*, 965 (1981)
572 N.N. Kavtaradze, N.P. Sokolova: Russ. J. Phys. Chem. *36*, 1529 (1962)
573 J. Pritchard: Trans. Faraday Soc. *59*, 437 (1963)
574 A.M. Bradshaw, J. Pritchard: Proc. Roy. Soc. Lond. *316A*, 169 (1970)
575 M.A. Chesters, J. Pritchard, M.L. Sims: "Infrared Reflection Spectra and Surface Potentials of Carbon Monoxide Chemisorbed on Copper, Silver, and Gold", in *Adsorption–Desorption Phenomena*, ed. by F. Ricca (Academic Press, London 1972) p. 277
576 M.L. Kottke, R.G. Greenler, H.G. Tompkins: Surf. Sci. *32*, 231 (1972)
577 G.W. Keulks, A. Ravi: J. Phys. Chem. *74*, 783 (1970)
578 H.A. Pearce: "Infrared Studies of Adsorption and Catalysis"; Ph.D. Thesis, University of East Anglia, Norwich (1974)
579 K.E. Hayes: Can. J. Spectrosc. *20*, 57 (1975)
580 G. McElhiney, H. Papp, J. Pritchard: Surf. Sci. *54*, 617 (1976)
581 J. Pritchard: "Infrared Reflection-Absorption Spectroscopy at Single Crystal Metal Surfaces", in Proc. Int. Conf. Vibrations in Adsorbed Layers, Jülich, June 1978, ed. by H. Ibach, S. Lehwald; Jül-Conf. 26, p.114
582 G.J. Slusser, N. Winograd: Surf. Sci. *95*, 53 (1980)
583 P. Hollins, J. Pritchard: "Reflection Absorption Infrared Spectroscopy: Application to Carbon Monoxide on Copper", in /17/, p. 125
584 R.P. Eischens, W.A. Pliskin, S.A. Francis: J. Chem. Phys. *22*, 1786 (1954)
585 R.P. Eischens, W.A. Pliskin: Adv. Catal. *10*, 2 (1958)
586 T.H. Wood, M.V. Klein, D.A. Zwemer: Surf. Sci. *107*, 625 (1981)
587 I. Pockrand, A. Otto: to be published
588 O. Ertürk, I. Pockrand, A. Otto: to be published
589 R.P. Eischens, S.A. Francis, W.A. Pliskin: J. Phys. Chem. *60*, 194 (1956)
590 R.A. Shigeishi, D.A. King: Surf. Sci. *58*, 379 (1976)
591 A.M. Bradshaw, F.M. Hoffmann: Surf. Sci. *72*, 513 (1978)
592 W. Erley, H. Wagner, H. Ibach: Surf. Sci. *80*, 612 (1979)
593 J. Pritchard, T. Catterick, R.K. Gupta: Surf. Sci. *53*, 1 (1975)
594 R.M. Hammaker, S.A. Francis, R.P. Eischens: Spectrochim. Acta *21*, 1295 (1965)
595 M. Moskovits, J.E. Hulse: Surf. Sci. *78*, 397 (1978)
596 M. Scheffler: Surf. Sci. *81*, 562 (1979)

597 P. Hollins: Surf. Sci. *107*, 75 (1981)
598 S. Efrima, H. Metiu: Surf. Sci. *109*, 109 (1981)
599 D.P. Woodruff, B.E. Hayden, K. Prince, A.M. Bradshaw: Surf. Sci. *123*, 397 (1982)
600 J.G. Roth, M.J. Dignam: Can. J. Chem. *54*, 1388 (1976)
601 S. Andersson: Surf. Sci. *89*, 477 (1979)
602 S. Andersson, B.N.J. Persson: Phys. Rev. Lett. *45*, 1421 (1980)
603 N.V. Richardson, A.M. Bradshaw: "Localized Vibrational Modes of Adsorbed Species", in Proc. Int. Conf. Vibrations in Adsorbed Layers, Jülich, June 1978, ed. by H. Ibach, S. Lehwald; Jül - Conf. 26, p. 2
604 A. Spitzer, H. Lüth: Surf. Sci. *102*, 29 (1981)
605 I. Pockrand: unpublished
606 C. Pettenkofer, I. Pockrand: unpublished
607 J.C. Tsang, J.E. Demuth, P.N. Sanda, J.R. Kirtley: Chem. Phys. Lett. *76*, 54 (1980)
608 F. Tuinstra, J.L. Koenig: J. Chem. Phys. *53*, 1126 (1970)
609 N. Wada, S.A. Solin, J.E. Wing: in *Physics of Semiconductors*, International Conference Series No. 43 (Institute of Physics, London 1979) p. 721
610 L.V. Del Priore, C. Doyle, J.D. Andrade: Appl. Spectrosc. *36*, 69 (1982)
611 P.H. Krupenie: J. Phys. Chem. Ref. Data *1*, 423 (1972)
612 J.A. Connor, E.A.V. Ebsworth: Adv. Inorg. Chem. *6*, 279 (1964)
613 V.J. Choy, C.J. O'Connor: Coord. Chem. Rev. *9*, 145 (1972)
614 L. Vaska: Acc. Chem. Res. *9*, 175 (1976)
615 D.M. Adams: *Metal - Ligand and Related Vibration* (Arnold, London 1967)
616 J. Weidlein, U. Müller, K. Dehnicke: *Schwingungsspektroskopie* (G. Thieme, Stuttgart 1982)
617 D.E. Tevault, R.A. deMarco, R.R. Smardzewski: J. Chem. Phys. *75*, 4168 (1981)
618 D.E. Tevault, R.L. Mowery, R.A. deMarco, R.R. Smardzewski: J. Chem. Phys. *74*, 4342 (1981)
619 U. Uhler: Ark. Fys. *7*, 125 (1954)
620 O. Appelblad, A. Lagerqvist: Phys. Scr. *10*, 307 (1974)
621 T. Shibahara, M. Mori: Bull. Chem. Soc. Jap. *51*, 1374 (1978)
622 D. McIntosh, G.A. Ozin: Inorg. Chem. *16*, 59 (1977)
623 D.E. Tevault, R.R. Smardzewski, M.W. Urban, K. Nakamoto: "Catalytic Intermediates in the $Ag-O_2$ System. Evidence for a non-Symmetric AgO_2 Molecule" to be published
624 D. McIntosh, G.A. Ozin: Inorg. Chem. *15*, 2869 (1976)
625 J.H. Darling, M.B. Garton-Sprenger, J.S. Ogden: Symp. Faraday Soc. *8*, 75 (1973)
626 H. Froitzheim, H. Ibach, S. Lehwald: Phys. Rev. *B14*, 1362 (1976)
627 S. Andersson: Sol. State Commun. *20*, 229 (1976)
628 G. Dalmai-Imelik, J.C. Bertolini, J. Rousseau: Surf. Sci. *63*, 67 (1977)
629 G.E. Thomas, W.H. Weinberg: J. Chem. Phys. *69*, 3611 (1978)
630 B.A. Sexton: J. Vac. Sci. Technol. *16*, 1033 (1979)
631 J.L. Gland, B.A. Sexton, G.B. Fisher: Surf. Sci. *95*, 587 (1980)
632 H. Steininger, S. Lehwald, H. Ibach: Surf. Sci. *123*, 1 (1982)
633 C. Backx, C.P.M. DeGroot, P. Biloen: Surf. Sci. *104*, 300 (1981)
634 B.A. Sexton, R.J. Madix: Chem. Phys. Lett. *76*, 294 (1980)
635 W. Erley, H. Ibach: Sol. State Commun. *37*, 937 (1981)
636 P. Hofmann, K. Horn, A.M. Bradshaw, K. Jacobi: Surf. Sci. *82*, L 610 (1979)
637 C. Leung, R. Gomer: Surf. Sci. *59*, 638 (1976)
638 D. Schmeisser, K. Jacobi: Surf. Sci. *108*, 421 (1981)
639 D. Schmeisser, J.E. Demuth, Ph. Avouris: Phys. Rev. *B26*, 4857 (1982)
640 P.R. Norton: Surf. Sci. *47*, 98 (1975)
641 M.A. Barteau, R.J. Madix: Surf. Sci. *97*, 101 (1980)
642 C.N.R. Rao, P. Vishnu Kamath, S. Yoshonath: Chem. Phys. Lett. *88*, 13 (1982)
643 J.F. Wendelken: Surf. Sci. *108*, 605 (1981)
644 A. Spitzer, H. Lüth: Surf. Sci. *118*, 121 (1982)
645 A. Spitzer, H. Lüth: Surf. Sci. *118*, 136 (1982)
646 B.A. Sexton: Surf. Sci. *88*, 299 (1979)
647 L.H. Dubois: Surf. Sci. *119*, 399 (1982)

648 P.A. Kilty, N.C. Rol, W.M.H. Sachtler: "Identification of Oxygen Complexes Adsorbed on Silver and their Function in the Catalytic Oxidation of Ethylene", in Proc. 5th Intern. Conf. Catal., ed. by J.W. Hightower, Miami, 1972; p. 64
649 A.W. Czanderna, S.C. Chen, J.R. Biegen: J. Catal. *33*, 163 (1974)
650 G. Rovida, F. Pratesi, M. Maglietta, E. Ferroni: Surf. Sci. *43*, 230 (1974)
651 H.A. Engelhardt, D. Menzel: Surf. Sci. *57*, 591 (1976)
652 H. Albers, W.J.J. van der Wal, G.A. Bootsma: Surf. Sci. *68*, 47 (1977)
653 J.E. Demuth, D. Schmeisser, Ph. Avouris: Phys. Rev. Lett. *47*, 1166 (1981)
654 C. Pettenkofer, I. Pockrand, A. Otto: Surf. Sci. *135*, 52 (1983)
655 C. Pettenkofer: unpublished
656 I. Pockrand: unpublished
657 R.C. Spiker Jr., L. Andrews: J. Chem. Phys. *59*, 1851 (1973)
658 L. Andrews, R.C. Spiker Jr.: J. Chem. Phys. *59*, 1863 (1973)
659 M. Bowker, M.A. Barteau, R.J. Madix: Surf. Sci. *92*, 528 (1980)
660 M.A. Barteau, R.J. Madix: J. Chem. Phys. *74*, 4144 (1981)
661 E.M. Stuve, R.J. Madix, B.A. Sexton: Chem. Phys. Lett. *89*, 48 (1982)
662 E.L. Force, A.T. Bell: J. Catal. *38*, 440 (1975)
663 K. Lochner, B. Reimer, H. Bässler: Chem. Phys. Lett. *41*, 388 (1976)
664 D. Bloor, W. Herschel, D.N. Batchelder: Chem. Phys. Lett. *45*, 411 (1977)
665 J.O'M. Bockris, A.K.N. Reddy: *Modern Electrochemistry*, Vol. 2 (Plenum, New York 1970) p. 739
666 M. Fleischmann, P.J. Hendra, I.R. Hill, M.E. Pemble: J. Electroanal. Chem. *117*, 243 (1981)
667 B. Pettinger, M.R. Philpott, J.G. Gordon II: J. Chem. Phys. *74*, 934 (1981)
668 B. Pettinger, M.R. Philpott, J.G. Gordon II: Surf. Sci. *105*, 469 (1981)
669 T.T. Chen, J.F. Owen, R.K. Chang, B.L. Laube: Chem. Phys. Lett. *89*, 356 (1982)
670 S.H. Macomber, T.E. Furtak, T.M. Devine: Surf. Sci. *122*, 556 (1982)
671 D. Eisenberg, W. Kauzmann: *The Structure and Properties of Water* (Oxford Univ. Press, London 1969)
672 G.C. Pimentel, A.L. McClellan: *The Hydrogen Bond* (Reinhold, New York 1960)
673 F. Franks: "The Properties of Ice", in *Water, a Comprehensive Treatise*, ed. by F. Franks, Vol. 1 (Plenum, New York 1972) p. 115
674 G.E. Walrafen: "Raman and Infrared Spectral Investigations of Water Structure", in *Water, a Comprehensive Treatise*, ed. by F. Franks, Vol. 1 (Plenum, New York 1972) p. 151
675 J.R. Scherer: "The Vibrational Spectroscopy of Water", in Advances in Infrared and Raman Spectroscopy, ed. by R.J.H. Clark and R.E. Hester, Vol. 5 (Heyden, London 1978) p. 149
676 M.G. Sceats, S.A. Rice: "Amorphous Solid Water and Its Relationship to Liquid Water: A Random Network Model for Water", in *Water, a Comprehensive Treatise*, ed. by F. Franks, Vol. 7 (Plenum, New York 1982) p. 83
677 H. Ibach, S. Lehwald: Surf. Sci. *91*, 435 (1980)
678 B.A. Sexton: Surf. Sci. *94*, 435 (1980)
679 G.B. Fisher, J.L. Gland: Surf. Sci. *94*, 446 (1980)
680 G.B. Fisher, B.A. Sexton: Phys. Rev. Lett. *44*, 683 (1980)
681 P.A. Thiel, F.M. Hoffmann, W.H. Weinberg: J. Chem. Phys. *75*, 5556 (1981)
682 K. Kretzschmar, J.K. Sass, A.M. Bradshaw: Surf. Sci. *115*, 183 (1982)
683 D.L. Doering, T.E. Madey: Surf. Sci. *123*, 305 (1982)
684 C. Benndorf, C. Nöbl, M. Rüsenberg, F. Thieme: Surf. Sci. *111*, 87 (1981)
685 C. Benndorf, C. Nöbl, F. Thieme: Surf. Sci. *121*, 249 (1982)
686 E.M. Stuve, R.J. Madix, B.A. Sexton: Surf. Sci. *111*, 11 (1981)
687 E.M. Stuve, R.J. Madix, B.A. Sexton: J. Vac. Sci. Technol. *20*, 590 (1982)
688 C. Au, J. Breza, M.W. Roberts: Chem. Phys. Lett. *66*, 340 (1979)
689 A.M. Baro, W. Erley: J. Vac. Sci. Technol. *20*, 580 (1982)
690 A. Spitzer, H. Lüth: Surf. Sci. *120*, 376 (1982)
691 I. Pockrand: Surf. Sci. *122*, L 569 (1982)
692 A.H. Hardin, K.B. Harvey: Spectrochim. Acta *29A*, 1139 (1973)
693 C.G. Venkatesh, S.A. Rice, J.B. Bates: J. Chem. Phys. *63*, 1065 (1975)
694 J.E. Bertie, E. Whalley: J. Chem. Phys. *40*, 1637 (1964)
695 M.J. Taylor, E. Whalley: J. Chem. Phys. *40*, 1660 (1964)

696 J.G. Bergman, J.P. Heritage, A. Pinczuk, J.M. Worlock, J.H. McFree: Chem. Phys. Lett. *68*, 412 (1979)
697 G. Laufer, J.T. Huneke, T.F. Schaaf: Chem. Phys. Lett. *82*, 571 (1981)
698 R.M. Hart, J.G. Bergman, A. Wokaun: Opt. Lett. *7*, 105 (1982)
699 K. Metcalfe, R.E. Hester: Chem. Phys. Lett. *94*, 411 (1983)
700 H. Kuhn, D. Möbius, H. Bücher: "Spectroscopy of Monolayer Assemblies", in *Physical Methods of Chemistry*, ed. by A. Weissberger, B.W. Rossiter, Vol. 1, Part IIIB (Wiley, New York 1972) p. 577
701 W. Knoll, M.R. Philpott, W.G. Golden: J. Chem. Phys. *77*, 219 (1982)
702 W. Knoll, M.R. Philpott, J.D. Swalen, A. Girlando: J. Chem. Phys. *77*, 2254 (1982)
703 R. Aroca, R.O. Loutfy: J. Raman Spectrosc. *12*, 262 (1982)
704 M. Moskovits, D.P. DiLella: J. Chem. Phys. *77*, 1655 (1982)
705 B.N.J. Persson: J. Phys. C: Solid State Phys. *11*, 4251 (1978)
706 R. Ruppin: J. Chem. Phys. *76*, 1681 (1982)
707 D.A. Weitz, S. Garoff, C.D. Hanson, T.J. Gramila, J.I. Gersten: Opt. Lett. *7*, 89 (1982)
708 J.A. Nimmo, D.H. Brown, W.E. Smith: Chem. Phys. Lett. *92*, 212 (1982)
709 H. Seki: J. Electron Spectrosc. Relat. Phenom. *30*, 287 (1983)
710 H. Seki: J. Vac. Sci. Technol. *20*, 584 (1982)
711 C.J. Sandroff, D.R. Herschbach: J. Phys. Chem. *86*, 3277 (1982)
712 H. Czichos: *Tribology, A Systems Approach to the Science and Technology of Friction Lubrication and Wear* (Elsevier, New York 1978)
713 T. Sakurai: J. Lubr. Technol. *103*, 473 (1981)
714 P.B. Dorain, K.U. von Raben, R.K. Chang, B.L. Laube: Chem. Phys. Lett. *84*, 405 (1981)
715 K.U. von Raben, R.K. Chang, P.B. Dorain, B.L. Laube: "Adsorbed Sulfur and Nitrogen Oxides on Silver Powders Detected by Surface Enhanced Raman Scattering", in Proc. Conf. on Lasers as Reactants and Probes in Chemistry, Washington, 1982; to be published
716 F.Kh. Ibragimov, V.M. Lysikov, N.Kh. Valitov, K.Kh. Khaziev: Russian J. Phys. Chem. *52*, 554 (1978)
717 H. Morawitz, M.R. Philpott: Phys. Rev. *B10*, 4863 (1974)
718 I. Pockrand, A. Brillante, D. Möbius: Chem. Phys. Lett. *69*, 499 (1980)
719 G. Ritchie, E. Burstein: Phys. Rev. *B24*, 4843 (1981)
720 A. Nitzan, L.E. Brus: J. Chem. Phys. *75*, 2205 (1981)
721 S. Garoff, D.A. Weitz, M.S. Alverez: Chem. Phys. Lett. *93*, 283 (1982)
722 G.M. Goncher, C.B. Harris: J. Chem. Phys. *77*, 3767 (1982)
723 A. Hjortsberg, W.P. Chen, E. Burstein, M. Pomerantz: Opt. Commun. *25*, 65 (1978)
724 H. Ueba, S. Ichimura: Phys. Stat. Sol. (b) *110*, K 161 (1982)
725 H. Ueba, S. Ichimura: Surf. Sci. *118*, L 273 (1982)
726 A. Hartstein, J.R. Kirtley, J.C. Tsang: Phys. Rev. Lett. *45*, 201 (1980)
727 A. Hatta, T. Ohshima, W. Suëtaka: Appl. Phys. *A29*, 71 (1982)
728 N. Bloembergen: *Nonlinear Optics* (Benjamin, New York 1964)
729 Y.R. Shen, C.K. Chen, T.F. Heinz, D. Ricard: "Surface-Enhanced Nonlinear Optical Effects and Detection of Adsorbed Molecular Monolayers", in *Laser Spectroscopy V*, ed. by A.R.W. McKellar, T. Oka, B.P. Stoicheff, Springer Series in Optical Sciences, Vol. 40 (Springer, Berlin, Heidelberg, New York 1981) p. 412
730 J.P. Heritage, A.M. Glass: "Nonlinear Optical Effects", in /67/, p. 391
731 H.J. Simon, D.E. Mitchell, J.G. Watson: Phys. Rev. Lett. *33*, 1531 (1974)
732 C.K. Chen, A.R.B. de Castro, Y.R. Shen: Opt. Lett. *4*, 393 (1979)
733 C.K. Chen, A.R.B. de Castro, Y.R. Shen: Phys. Rev. Lett. *46*, 145 (1981)
734 A. Wokaun, J.G. Bergman, J.P. Heritage, A.M. Glass, P.F. Liao, D.H. Olson: Phys. Rev. *B24*, 849 (1981)
735 T.F. Heinz, C.K. Chen, D. Ricard, Y.R. Shen: Chem. Phys. Lett. *83*, 180 (1981)
736 G.S. Agarwal, S.S. Jha: Sol. State Commun. *41*, 499 (1982)

737 C.K. Chen, A.R.B. de Castro, Y.R. Shen, F. De Martini: Phys. Rev. Lett. *43*, 946 (1979)
738 H. Chew, D.-S. Wang, M. Kerker: J. Opt. Soc. Am. *B1*, 56 (1984)
739 D.V. Murphy, K.U. von Raben, R.K. Chang, P.B. Dorain: Chem. Phys. Lett. *85*, 43 (1982)
740 A.M. Glass, A. Wokaun, J.P. Heritage, J.G. Bergman, P.F. Liao, D.H. Olson: Phys. Rev. *B24*, 4906 (1981)
741 C.K. Chen, T.F. Heinz, D. Ricard, Y.R. Shen: Phys. Rev. Lett. *46*, 1010 (1981)
742 H. Seki, T.J. Chuang: "The Detection by SERS of Resonantly Excited Desorption of Pyridine from Silver Island Films by IR Laser Absorption", to be published
743 T.J. Chuang, H. Seki: Phys. Rev. Lett. *49*, 382 (1982)
744 D.L. Allara, C.A. Murray, S. Bodoff: Bull. Am. Phys. Soc. *26*, 338 (1981)
745 C.A. Murray: J. Electron Spectrosc. Relat. Phenom. *29*, 371 (1983)
746 F.M. Hoffmann: Surf. Sci. Rep. *3*, 107 (1983)
747 W.G. Golden, D.D. Saperstein, M.W. Severson, J. Overend: J. Phys. Chem. *88*, 574 (1984)
748 T.E. Furtak: J. Electroanal. Chem. *150*, 375 (1983)
749 F.R. Aussenegg, A. Leitner, M.E. Lippitsch (eds.): *Surface Studies with Lasers*, Springer Series in Chemical Physics, Vol. 33 (Springer, Berlin, Heidelberg, New York 1983)
750 A. Aurengo, Y. Levy, R. Dupeyrat: Appl. Opt. *22*, 602 (1983)
751 Ph. Avouris, J.C. Tsang, J.R. Kirtley: J. Vac. Sci. Technol. *A1*, 1240 (1983)
752 A. Campion, D.R. Mullins: "Normal (Unenhanced) Raman Scattering from Pyridine Adsorbed on the Low-Index Faces of Silver", in /749/, p. 36
753 A. Campion, V.M. Grizzle, D.R. Mullins, J.K. Brown: J. Physique *44*, C10 - 341 (1983)
754 D.R. Mullins, A. Campion: J. Phys. Chem. *88*, 8 (1984)
755 R.E. Benner, K.U. von Raben, K.C. Lee, J.F. Owen, R.K. Chang, B.L. Laube: Chem. Phys. Lett. *96*, 65 (1983)
756 J.E. Potts, R. Merlin, D.L. Partin: Phys. Rev. *B27*, 3905 (1983)
757 M. Fleischmann, P.R. Graves, I.R. Hill, J. Robinson: Chem. Phys. Lett. *95*, 322 (1983)
758 D.V. Murphy, S.R.J. Brueck: Opt. Lett. *8*, 494 (1983)
759 T. Springer: "Investigation of Vibrations in Metal Hydrides by Neutron Spectroscopy", in *Hydrogen in Metals I*, ed. by G. Alefeld and J. Völkl, Topics in Applied Physics, Vol. 28 (Springer, Berlin, Heidelberg, New York 1978) p. 75
760 E. Wicke, A. Brodowsky: "Hydrogen in Palladium and Palladium Alloys", in *Hydrogen in Metals II*, ed. by G. Alefeld and J. Völkl, Topics in Applied Physics, Vol. 29 (Springer, Berlin, Heidelberg, New York 1978) p. 73
761 K. Ohtaka, M. Inoue: "Enhanced Raman Scattering from a Dielectric Sphere", in /749/, p. 97
762 K. Ohtaka, M. Inoue: "Enhanced Raman Scattering from a Periodic Monolayer of Dielectric Spheres", in /749/, p. 101
763 C. Pettenkofer, A. Otto: J. Physique *44*, C10 - 341 (1983)
764 H. Metiu, P. Das: "The Electromagnetic Theory of Surface Enhanced Spectroscopy", to be published in Ann. Rev. Phys. Chem. (1984)
765 D. Sarid: Phys. Rev. Lett. *47*, 1927 (1981)
766 D. Sarid, R.T. Deck, A.E. Craig, R.K. Hickernell, R.S. Jameson, J.J. Fasano: Appl. Opt. *21*, 3993 (1982)
767 P. Sheng, R.S. Stepleman, P.N. Sanda: Phys. Rev. *B26*, 2907 (1982)
768 M. Nevière, R. Reinisch: Phys. Rev. *B26*, 5403 (1982)
769 R. Reinisch, M. Nevière: J. Optics (Paris) *13*, 81 (1982)
770 M. Weber, D.L. Mills: Phys. Rev. *B27*, 2698 (1983)
771 M. Yamashita, M. Tsuji: J. Phys. Soc. Jap. *52*, 2462 (1983)
772 N.E. Glass, A.A. Maradudin, V. Celli: J. Opt. Soc. Am. *73*, 1240 (1983)
773 A. Wirgin, T. Lôpez-Rios: Opt. Commun. *48*, 416 (1984)
774 P.K. Aravind, H. Metiu: Surf. Sci. *124*, 506 (1983)
775 R. Ruppin: Surf. Sci. *127*, 108 (1983)

776 P.W. Barber, R.K. Chang, H. Massoudi: Phys. Rev. *B27*, 7251 (1983)
777 W.A. Kraus, G.C. Schatz: Chem. Phys. Lett. *99*, 353 (1983)
778 M. Kerker, D.S. Wang: Chem. Phys. Lett. *104*, 516 (1984)
779 G.S. Agarwal, S.S. Jha: Phys. Rev. *B26*, 4013 (1982)
780 K. Arya, R. Zeyher: Phys. Rev. *B28*, 4090 (1983)
781 E.V. Albano, S. Daiser, G. Ertl, R. Miranda, K. Wandelt, N. Garcia: Phys. Rev. Lett. *51*, 2314 (1983)
782 H. Seki, T.J. Chuang: Chem. Phys. Lett. *100*, 393 (1983)
783 C. Pettenkofer, O. Ertürk, A. Otto: "On the Contribution of Electromagnetic Enhancement to SERS of Cold Deposited Silver and Copper Films", to be published
784 S.A. Lyon, J.M. Worlock: Phys. Rev. Lett. *51*, 593 (1983)
785 A. Otto, J. Billmann, J. Eickmanns, O. Ertürk, C. Pettenkofer: Surf. Sci. *138*, 319 (1984)
786 M.E. Lippitsch, F.R. Aussenegg: "The Charge Transfer Contribution to Surface Enhanced Raman Scattering", in /749/, p. 41
787 M.E. Lippitsch: Phys. Rev. *B29*, 3104 (1984)
788 H. Ueba: Surf. Sci. *131*, 328 (1983)
789 H. Ueba: Surf. Sci. *133*, L 432 (1983)
790 J.F. Brazdil, E.B. Yeager: J. Phys. Chem. *85*, 995 (1981)
791 J.F. Brazdil, E.B. Yeager: J. Phys. Chem. *85*, 1005 (1981)
792 J.F. Brazdil, E.B. Yeager: J. Phys. Chem. *85*, 2194 (1981)
793 H. Ueba: Surf. Sci. *129*, L 267 (1983)
794 Ph. Avouris, J.E. Demuth: "Electronically Excited States of Adsorbates on Metal Surfaces", in /749/, p. 24
795 J. Billmann, A. Otto: Sol. State Commun. *44*, 105 (1982)
796 A. Otto: J. Electron Spectr. Relat. Phenom. *29*, 329 (1983)
797 M.L.A. Temperini, W.J. Barreto, O. Sala: Chem. Phys. Lett. *99*, 148 (1983)
798 T.E. Furtak, S.H. Macomber: Chem. Phys. Lett. *95*, 328 (1983)
799 J.R. Lombardi, R.L. Birke, L.A. Sanchez, I. Bernard, S.C. Sun: Chem. Phys. Lett. *104*, 240 (1983)
800 J. Billmann, A. Otto: Surf. Sci. *138*, 1 (1984)
801 J. Thietke, J. Billmann, A. Otto: "Charge Transfer Excitations in SERS: Comparative Study of Benzene, Pyridine and Pyrazine", to be published
802 A. Otto: Phys. Rev. *B27*, 5132 (1983)
803 V.V. Marinyuk, R.M. Lazorenko-Manevich, Ya.M. Kolotyrkin: Sol. State Commun. *43*, 721 (1982)
804 T. Watanabe, N. Yanagihara, K. Honda, B. Pettinger, L. Moerl: Chem. Phys. Lett. *96*, 649 (1983)
805 J.F. Owen, T.T. Chen, R.K. Chang, B.L. Laube: Surf. Sci. *131*, 195 (1983)
806 T.E. Furtak, D. Roy: Phys. Rev. Lett. *50*, 1301 (1983)
807 T. Watanabe, O. Kawanami, K. Honda, B. Pettinger: Chem. Phys. Lett. *102*, 565 (1983)
808 J.F. Owen, R.K. Chang: Chem. Phys. Lett. *104*, 59 (1984)
809 T.H. Wood: Phys. Rev. *B27*, 5137 (1983)
810 Y. Mo, I. Mörke, P. Wachter: Surf. Sci. *133*, L 452 (1983)
811 P.K.K. Pandey, G.C. Schatz: J. Chem. Phys. *80*, 2959 (1984)
812 S.H. Macomber, T.E. Furtak, T.M. Devine: Chem. Phys. Lett. *90*, 439 (1982)
813 T.T. Chen, K.U. von Raben, J.F. Owen, R.K. Chang, B.L. Laube: Chem. Phys. Lett. *91*, 494 (1982)
814 F. Barz, J.G. Gordon II, M.R. Philpott, M.J. Weaver: Chem. Phys. Lett. *91*, 291 (1982)
815 A. Regis, P. Dumas, J. Corset: "Contributions of Charge Transfer Complexes and Photochemical Effects to SERS at a Silver Electrode", in /749/, p. 50
816 T.P. Mernagh, R.P. Cooney: J. Raman Spectr. *14*, 138 (1983)
817 M.R. Mahoney, R.P. Cooney: J. Phys. Chem. *87*, 4589 (1983)
818 R.P. Cooney, T.P. Mernagh, M.R. Mahoney, J.A. Spink: J. Phys. Chem. *87*, 5314 (1983)
819 P.H. McBreen, M. Moskovits: J. Appl. Phys. *54*, 329 (1983)
820 T. Lôpez-Rios, Y. Borensztein, G. Vuye: J. Physique *44*, C10 - 353 (1983)
821 Y. Borensztein, T. Lôpez-Rios, G. Vuye: J. Physique *44*, C10 - 475 (1983)
822 T. Lôpez-Rios, Y. Borensztein, B. Vuye: J. Physique *44*, L 99 (1983)

823 J. Eickmanns, A. Goldmann, A. Otto: "The Structure of Cold Deposited Ag Films Investigated with UPS, XPS, and TDS of Xe and Oxygen", to be published
824 T. López-Rios, G. Vuye, Y. Borensztein: Surf. Sci. *131*, L 367 (1983)
825 T. Yamaguchi, M. Ogawa, H. Takahashi, N. Saito: Surf. Sci. *129*, 232 (1983)
826 A. Ljunbert, P. Apell: Sol. State Commun. *46*, 47 (1983)
827 W. Ekardt, D.B. Tran Thoai, F. Frank, W. Schulze: Sol. State Commun. *46*, 571 (1983)
828 G.A. Ozin, S.A. Mitchell, D.F. McIntosh, S.M. Mattar, J. Garcia-Prieto: J. Phys. Chem. *87*, 4651 (1983)
829 G.A. Ozin, S.A. Mitchell, S.M. Mattar, J. Garcia-Prieto: J. Phys. Chem. *87*, 4666 (1983)
830 D.M. Kolb, H.H. Rotermund, W. Schrittenlacher, W. Schroeder: J. Chem. Phys. *80*, 695 (1984)
831 T. Okada, T. Iwaki, K. Yamamoto, H. Kasahara, K. Abe: Sol. State Commun. *49*, 809 (1984)
832 G.A. Ozin, S.A. Mitchell: J. Phys. Chem. (1984), in press
833 F.P. Netzer, J.-U. Mack: Chem. Phys. Lett. *95*, 492 (1983)
834 J.A. Creighton, M.S. Alvarez, D.A. Weitz, S. Garoff, M.W. Kim: J. Phys. Chem. *87*, 4793 (1983)
835 H. Yamada, Y. Yamamoto: Surf. Sci. *134*, 71 (1983)
836 C. Minot, M.A. van Howe, G.A. Somorjai: Surf. Sci. *127*, 441 (1982)
837 R.J. Koestner, M.A. van Howe, G.A. Somorjai: J. Phys. Chem. *87*, 203 (1983)
838 T.M. Gentle, E.L. Muetterties: J. Phys. Chem. *87*, 2469 (1983)
839 A.B. Anderson, S.P. Mehandru: Surf. Sci. *136*, 398 (1984)
840 W. Hasse, H.-L. Günter, M. Henzler: Surf. Sci. *126*, 479 (1983)
841 N. Freyer, G. Pirug, H.P. Bonzel: Surf. Sci. *126*, 487 (1983)
842 N. Freyer, G. Pirug, H.P. Bonzel: Surf. Sci. *125*, 327 (1983)
843 I. Ratajczykowa, I. Szymerska: Chem. Phys. Lett. *96*, 243 (1983)
844 C. Backx, C.P.M. de Groot, P. Biloen, W.M.H. Sachtler: Surf. Sci. *128*, 81 (1983)
845 D.R. Lloyd, F.P. Netzer: Surf. Sci. *129*, L 249 (1983)
846 J.R. Creighton, J.M. White: Surf. Sci. *129*, 327 (1983)
847 E. Schmiedl, P. Wissmann, E. Wittmann: Surf. Sci. *135*, 341 (1983)
848 R.J. Koestner, J. Stöhr, J.L. Gland, J.A. Horsley: Chem. Phys. Lett. *105*, 332 (1984)
849 J.R. Creighton, K.M. Ogle, J.M. White: Surf. Sci. *138*, L 137 (1984)
850 P.D. Szuromi, W.H. Weinberg: J. Vac. Sci. Technol. *A1*, 1219 (1983)
851 P.D. Szuromi, J.R. Engstrom, W.H. Weinberg: J. Chem. Phys. *80*, 508 (1984)
852 M. Stoukides, C.G. Vayenas: J. Catal. *82*, 45 (1983)
853 N.J. DiNardo, J.E. Demuth, Ph. Avouris: J. Vac. Sci. Technol. *A1*, 1244 (1983)
854 W. Sesselmann, B. Woratschek, G. Ertl, J. Küppers, H. Haberland: Surf. Sci. *130*, 245 (1983)
855 N.J. DiNardo, J.E. Demuth, Ph. Avouris: Phys. Rev. *B27*, 5832 (1983)
856 L.L. Kesmodel: J. Chem. Phys. *79*, 4646 (1983)
857 A.V. Bobrov, A.N. Gass, O.I. Kapusta, N.M. Omel'yanovskaya: J. Physique *44*, C10 - 327 (1983)
858 J.S. Suh, D.P. DiLella, M. Moskovits: J. Phys. Chem. *87*, 1540 (1983)
859 P.S. Bagus, C.J. Nelin, C.W. Bauschlicher, Jr.: Phys. Rev. *B28*, 5423 (1983)
860 B.E. Nieuwenhuys: Surf. Sci. *126*, 307 (1983)
861 J.L. Gland, M.R. McClellan, F.R. McFeely: J. Chem. Phys. *79*, 6349 (1983)
862 J.L. Gland, M.R. McClellan, F.R. McFeely: J. Vac. Sci. Technol. *A1*, 1070 (1983)
863 B.E. Hayden, A.M. Bradshaw: Surf. Sci. *125*, 787 (1983)
864 G.E. Mitchell, J.L. Gland, J.M. White: Surf. Sci. *131*, 167 (1983)
865 R. Ducros, B. Tardy, J.C. Bertolini: Surf. Sci. *128*, L 219 (1983)
866 M.W. Severson, W.J. Tornquist, J. Overend: J. Phys. Chem. *88*, 469 (1984)
867 P. Gelin, J.T. Yates, Jr.: Surf. Sci. *136*, L 1 (1984)
868 B.E. Nieuwenhuys, G.A. Kok: Thin Sol. Films *106*, L 95 (1983)
869 S. Chiang, R.G. Tobin, P.L. Richards, P.A. Thiel: Phys. Rev. Lett. *52*, 648 (1984)
870 S. Efrima, H. Metiu: Surf. Sci. *92*, 433 (1980)

871 S. Efrima: Surf. Sci. *114*, L 29 (1982)
872 P. Apell: Sol. State. Commun. *47*, 615 (1983)
873 S. Andersson: J. Vac. Sci. Technol. *A1*, 1242 (1983)
874 W. Eberhardt, E.W. Plummer: Phys. Rev. *B28*, 3605 (1983)
875 P. Hollins, J. Pritchard: Surf. Sci. *134*, 91 (1983)
876 C. Somerton, C.F. McConville, D.P. Woodroff, D.E. Grider, N.V. Richardson: Surf. Sci. *138*, 31 (1984)
877 P. Hollins, K.J. Davies, J. Pritchard: Surf. Sci. *138*, 75 (1984)
878 M. Watanabe, P. Wissmann: Surf. Sci. *138*, 95 (1984)
879 F. Frank, W. Schulze, B. Tesche, F.W. Froben: Chem. Phys. Lett. *103*, 336 (1984)
880 L. Papagno, L.S. Caputi, F. Ciccaci, C. Mariani: Surf. Sci. *128*, L 209 (1983)
881 A.D. van Langeveld, F.C.M.J.M. van Delft, V. Ponec: Surf. Sci. *135*, 93 (1983)
882 A.D. van Langeveld, F.C.M.J.M. van Delft, V. Ponec: Surf. Sci. *134*, 665 (1983)
883 J.N. Rouzaud, A. Oberlin, C. Beny-Bassez: Thin Sol. Films *105*, 75 (1983)
884 C. Nyberg, C.G. Tengstål: Surf. Sci. *126*, 163 (1983)
885 A.M. Barö, L. Ollé: Surf. Sci. *126*, 170 (1983)
886 C.W. Bauschlicher, Jr., P.S. Bagus: Phys. Rev. Lett. *52*, 200 (1984)
887 K.H. Rieder: Phys. Rev. *B27*, 6978 (1983)
888 R.J. Behm, C.R. Brundle: J. Vac. Sci. Technol. *A1*, 1223 (1983)
889 S. Daiser, K. Wandelt: Surf. Sci. *128*, L 213 (1983)
890 G.B. Fisher, S.J. Schmieg: J. Vac. Sci. Technol. *A1*, 1064 (1983)
891 N.R. Avery: Chem. Phys. Lett. *96*, 371 (1983)
892 A.J. Algra, E.P.Th.M. Suurmeijer, A.L. Boers: Surf. Sci. *128*, 207 (1983)
893 P.H. McBreen, M. Moskovits: J. Phys. Chem. *87*, 4843 (1983)
894 M. Ayyoob, M.S. Hedge: Surf. Sci. *133*, 516 (1983)
895 P. Vishnu Kamath, C.N.R. Rao: J. Phys. Chem. *88*, 464 (1984)
896 K.C. Prince, A.M. Bradshaw: Surf. Sci. *126*, 49 (1983)
897 M.A. Barteau, R.J. Madix: Chem. Phys. Lett. *97*, 85 (1983)
898 C. Benndorf, M. Franck, F. Thieme: Surf. Sci. *128*, 417 (1983)
899 R.L. Martin, P.J. Hay: Surf. Sci. *130*, L 283 (1983)
900 J.-H. Lin, B.J. Garrison: J. Chem. Phys. *80*, 2904 (1984)
901 R. Kötz, B.E. Hayden: Surf. Sci. *135*, 374 (1983)
902 S.M. Sonchik, L. Andrews, K.D. Carlson: J. Phys. Chem. *87*, 2004 (1983)
903 S.A. Rice, M.S. Bergren, A.C. Belch, G. Nielsson: J. Phys. Chem. *87*, 4295 (1983)
904 P.D. Schulze, S.L. Shaffer, R.L. Hance, D.L. Utley: J. Vac. Sci. Technol. *A1*, 97 (1983)
905 S. Ciraci, H. Wagner: Phys. Rev. *B27*, 5180 (1983)
906 D. Schmeisser, F.J. Himpsel, G. Hollinger: Phys. Rev. *B27*, 7813 (1983)
907 F.P. Netzer, T.E. Madey: Surf. Sci. *127*, L 102 (1983)
908 J.R. Creighton, J.M. White: Surf. Sci. *136*, 449 (1984)
909 K. Bange, D.E. Grider, T.E. Madey, J.K. Sass: Surf. Sci. *136*, 38 (1984)
910 M. Klaua, T.E. Madey: Surf. Sci. *136*, L 42 (1984)
911 S. Andersson, C. Nyberg, C.G. Tengstål: Chem. Phys. Lett. *104*, 305 (1984)
912 J.F. Owen, R.K. Chang: Chem. Phys. Lett. *104*, 510 (1984)
913 C.G. Blatchford, M. Kerker, D.-S. Wang: Chem. Phys. Lett. *100*, 230 (1983)
914 C. Pettenkofer, A. Otto: to be published
915 E. Burstein, G. Burns, F.H. Dacol: Sol. State Commun. *46*, 595 (1983)
916 K. Shoji, Y. Kobayashi, K. Itoh: Chem. Phys. Lett. *102*, 179 (1983)
917 N.T. Liang, T.T. Chen, H. Chang, Y.C. Chou, S. Wang: Opt. Lett. *8*, 374 (1983)
918 K.U. von Raben, P.B. Dorain, T.T. Chen, R.K. Chang: Chem. Phys. Lett. *95*, 269 (1983)
919 C.J. Sandroff, D.A. Weitz, J.C. Chung, D.R. Herschbach: J. Phys. Chem. *87*, 2127 (1983)
920 R.P. Van Duyne, J.P. Haushalter: J. Phys. Chem. *87*, 2999 (1983)
921 C.J. Sandroff, S. Garoff, K.P. Leung: Chem. Phys. Lett. *96*, 547 (1983)
922 T. Nanba, T.P. Martin: Phys. Stat. Sol. (a) *76*, 235 (1983)
923 A. Wokaun, H.-P. Lutz, A.P. King: "Distance Dependence in Surface Enhanced Luminescence", in /749/, p. 86
924 A. Wokaun, H.-P. Lutz, A.P. King, U.P. Wild, R.R. Ernst: J. Chem. Phys. *79*, 509 (1983)

925 A. Leitner, M.E. Lippitsch, F.R. Aussenegg: "Picosecond Fluorescence Decay
     of Dye Molecules Adsorbed to Small Metal Particles", in /749/, p. 90
926 A. Hatta, Y. Suzuki, W. Suëtaka: "Infrared Absorption Enhancement of Mono-
     layer Species on Thin Evaporated Ag Films by Use of the Kretschmann Configu-
     ration: Evidence for Two Types of Enhanced Surface Electric Fields", to be
     published
927 M.A. Chesters, O. Ertürk, A. Otto: "Enhanced Vibrational Excitation of Ad-
     sorbates Through Interaction with Metallic e-h-Pairs Observed by Transmission
     Infrared Spectroscopy", to be pusblished
928 R. Reinisch, M. Nevière: Phys. Rev. $B28$, 1870 (1983)
929 D.V. Murphy, K.U. von Raben, T.T. Chen, J.F. Owen, R.K. Chang: Surf. Sci.
     124, 529 (1983)
930 C.K. Chen, T.F. Heinz, D. Ricard, Y.R. Shen: Phys. Rev. $B27$, 1965 (1983)
931 D.S. Chemla, J.P. Heritage, P.F. Liao, E.D. Isaacs: Phys. Rev. $B27$, 4553
     (1983)
932 G.S. Agarwal, S.S. Jha: Phys. Rev. $B28$, 478 (1983)
933 P.F. Liao: "Surface Enhanced Optical Processes", in /749/, p. 72
934 T.F. Heinz, H.W.K. Tom, Y.R. Shen: "Nonlinear Optical Detection of Adsorbed
     Monolayers", in /749/, p. 79
935 M. Nevière: "Electromagnetic Resonances and Enhanced Nonlinear Optical
     Effects", in /749/, p. 94
936 J.C. Quail, H.J. Simon: J. Opt. Soc. Am. $B1$, 317 (1984)

Subject Index

106,124,130,132
Dehydrogenation 76,80,83,88
Density of states (phonons) 97
Depolarisation measurements 1,9
Desorption
  laser stimulated 78
  resonantly excited 123
Dielectric function 12,13,43
  effective 13
  particle size dependence 12,13,23,
    129
  spatial dispersion 14
Dipole-dipole coupling
  metal particles 13
  vibrating adsorbates 71,130
Disproportionation 77,82
Dynamical dipole moment 89

eh-pair excitation 132
Electrochemistry 4,107
Electrodes 5,6,7,9,20,26,48,127,129,
  131
Electromagnetic resonance 12,13,16,
  39,44,54,66,114,120,121,124,125,128
  inhomogeneous broadening 13,23
Electron transfer (donation) 89,101
Energy transfer, radiationless 132
Enhancement (SERS)
  adsorbate specifity 18
  mode specifity 9,18,27
Enhancement mechanisms 11ff
  chemical (molecular) 11,14ff,16,17,
    18,58,124,128,131
  electromagnetic (classical, local
    field) 11ff,15,16,17,18,44,124,
    125,128,132
  "field" 14,15,16,18,36,128
  "orchid" 16,17
Epoxidation 99
Excitation spectra (SERS) 9,18,30,
  39ff,45,66,128

Excitonic excitations 128

Fano resonance 67,71,132
Fermi resonance 110,131
Fresnel equations 8,9
Films
  coldly evaporated 4,7,20ff,54ff,101,
    124,128,129,130,131
  island 7,9,13,22ff,46,51,52,128,132
  vapour deposited 9,47,53,130
First layer effect, see surface en-
  hanced Raman scattering
Fischer-Tropsch synthesis 89
Fluorescence 115,120,132
  surface enhanced 120
  two photon, surface enhanced 122
Four-wave mixing, surface enhanced 132

Gas aggregation technique 25
Grating surfaces, see rough surfaces
Guided light modes 2,114,127

Hartree-Fock cluster calculations 15,
  129
Helmholtz layer 131
High temperature anneal 15,129
Hydrogenation, catalytic 59
Hydrogen bonding 107,108,110
  disruption 131
Hydrogenolysis 59
Hyper Raman scattering, surface en-
  hanced 122

Image dipole 13
Image field effects, see enhancement
  mechanisms
Infrared absorption 1
  surface enhanced 121,122,132
Interband transitions 127,128
Impurity lines, see surface enhanced
  Raman scattering

Material Index

Light Scattering in Solids IV

Electronic Scattering, Spin Effects, SERS and Morphic Effects

Editors: **M. Gardona, G. Güntherodt**
1984. 322 figures. Approx. 560 pages
(Topics in Applied Physics, Volume 54)
ISBN 3-540-11942-6

Contents: *M. Cardona, G. Güntherodt:* Introduction. - *A. Pinczuk, G. Abstreiter, M. Cardona:* Light Scattering by Free Carrier Excitations in Semiconductors. - *S. Geschwind, R. Romestain:* High Resolution Spin-Flip Raman-Scattering in CdS. - *G. Güntherodt, R. Zeyher:* Spin-Dependent Raman Scattering in Magnetic Semiconductors. - *G. Güntherodt, R. Merlin:* Raman Scattering in Rare-Earth Chalcogenides. - *A. Otto:* Surface Enhanced Raman Scattering: "Classical" and "Chemical" Origins. - *K. Arya, R. Zeyher:* Theory of Surface-Enhanced Raman Scattering. - *B. A. Weinstein, R. Zallen:* Pressure-Raman Effects in Covalent and Molecular Solids. - Errata for Light Scattering in Solids II (TAP 50). - Subject Index.

Chemistry and Physics of Solid Surfaces IV

Editors: **R. Vanselow, R. Howe**
1982. 247 figures. XIII, 496 pages
(Springer Series in Chemical Physics, Volume 20)
ISBN 3-540-11397-5

Contents: Development of Photoemission as a Tool for Surface Science: 1900–1980. - Auger Spectroscopy as a Probe of Valence Bonds and Bands. - SIMS of Reactive Surfaces. - Chemisorption Investigated by Ellipsometry. - The Implications for Surface Science of Doppler-Shift Laser Fluorescence Spectroscopy. - Analytical Electron Microscopy in Surface Science. - He Diffraction as a Probe of Semiconductor Surface Structures. - Studies of Adsorption at Well-Ordered Electrode Surfaces Using Low-Energy Electron Diffraction. - Low-Energy Electron Diffraction Studies of Physically Adsorbed Films. - Monte Carlo Simulations of Chemisorbed Overlayers. - Critical Phenomena of Chemisorbed Overlayers. - Structural Defects in Surfaces and Overlayers. - Some Theoretical Aspects of Metal Clusters, Surfaces, and Chemisorption. - The Inelastic Scattering of Low-Energy Electrons by Surface Excitations; Basic Mechanisms. - Electronic Aspects of Adsorption Rates. - Thermal Desorption. - Field Desorption and Photon-Induced Field Desorption. - Segregation and Ordering at Alloy Surfaces Studied by Low-Energy Ion Scattering. - The Effects of Internal Surface Chemistry on Metallurgical Properties. - Subject Index.

Chemistry and Physics of Solid Surfaces V

Editors: **R. Vanselow, R. Howe**
1984. 303 figures. Approx. 460 pages
(Springer-Series in Chemical Physics, Volume 35)
ISBN 3-540-13315-1

Vibrational Spectroscopy of Adsorbates

Editor: **R. F. Willis**
With contributions by numerous experts
1980. 97 figures, 8 tables. XII, 184 pages
(Springer Series in Chemical Physics, Volume 15)
ISBN 3-540-10429-1

Contents: Introduction. - Theory of Dipole Electron Scattering from Adsorbates. - Angle and Energy Dependent Electron Impact Vibrational Excitation of Adsorbates. - Adsorbate Induced Optical Phonons. - Inelastic Electron Tunnelling Spectroscopy. - Inelastic Molecular Beam Scattering from Surfaces. - Neutron Scattering Studies. - Reflection Absorption Infrared Spectroscopy: Application to Carbon Monoxide on Copper. - Raman Spectroscopy of Adsorbates at Metal Surfaces. - Vibrations of Monatomic and Diatomic Ligands in Metal Clusters and Complexes - Analogies with Vibrations of Adsorbed Species on Metals. - Coupling Induced Vibrational Frequency Shifts and Island Size Determination: CO on Pt {001} and Pt {111}.

Springer-Verlag Berlin Heidelberg New York Tokyo

Secondary Ion Mass Spectrometry SIMS-II

Proceedings of the Second International Conference on Secondary Ion Mass Spectrometry (SIMS II) Stanford University, Stanford, California, USA, August 27–31, 1979

Editors: **A. Benninghoven, C.A. Evans, Jr., R.A. Powell, R. Shimizu, H.A. Storms**

1979. 234 figures, 21 tables. XIII, 298 pages (Springer Series in Chemical Physics, Volume 9) ISBN 3-540-09843-7

Contents: Fundamentals. – Quantitation. – Semiconductors. – Static SIMS. – Metallurgy. – Instrumentation. – Geology. – Panel Discussion. – Biology. – Combined Techniques. – Postdeadline Papers.

Ion Formation from Organic Solids

Proceedings of the Second International Conference, Münster, Federal Republic of Germany, September 7–9, 1982

Editor: **A. Benninghoven**

1983. 170 figures. IX, 269 pages (Springer Series in Chemical Physics, Volume 25) ISBN 3-540-12244-3

Contents: Field Desorption. – ^{252}Cf-Plasma Desorption. – Secondary Ion Mass Sepctrometry (SIMS) Including FAB. – Laser Induced Ion Formation. – Other Ion Formation Processes. – Index of Contributors.

Secondary Ion Mass Spectrometry, SIMS IV

Proceedings of the Fourth International Conference on Secondary Ion Mass Spectrometry (SIMSIV), Osaka, Japan, November 13–19, 1983

Editors: **A. Benninghoven, J. Okano, R. Schimizu**

1984. 392 figures. Approx. 480 pages (Springer Series in Chemical Physics, Volume 36) ISBN 3-540-13316-X

Contents: Fundamentals. – Quantification. – Instrumentation. – Combined and Static SIMS. – Application to Semiconductor and Depth Profiling. – Organic SIMS. – Applications: Metallic and Inorganic Materials. – Geology. – Biology. – Index of Contributors.

Secondary Ion Mass Spectrometry SIMS III

Proceedings of the Third International Conference, Technical University, Budapest, Hungary, August 30–September 5, 1981

Editors: **A. Benninghoven, J. Giber, J. László, M. Riedel, H.W. Werner**

1982. 289 figures. XI, 444 pages (Springer Series in Chemical Physics, Volume 19 ISBN 3-540-11372-X

Contents: Instrumentation. – Fundamentals I. Ion Formation. – Fundamentals II. Depth Profiling. – Quantification. – Application I. Depth Profiling. – Application II. Surface Studies, Ion Microscopy. – Index of Contributors.

Springer-Verlag Berlin Heidelberg New York Tokyo